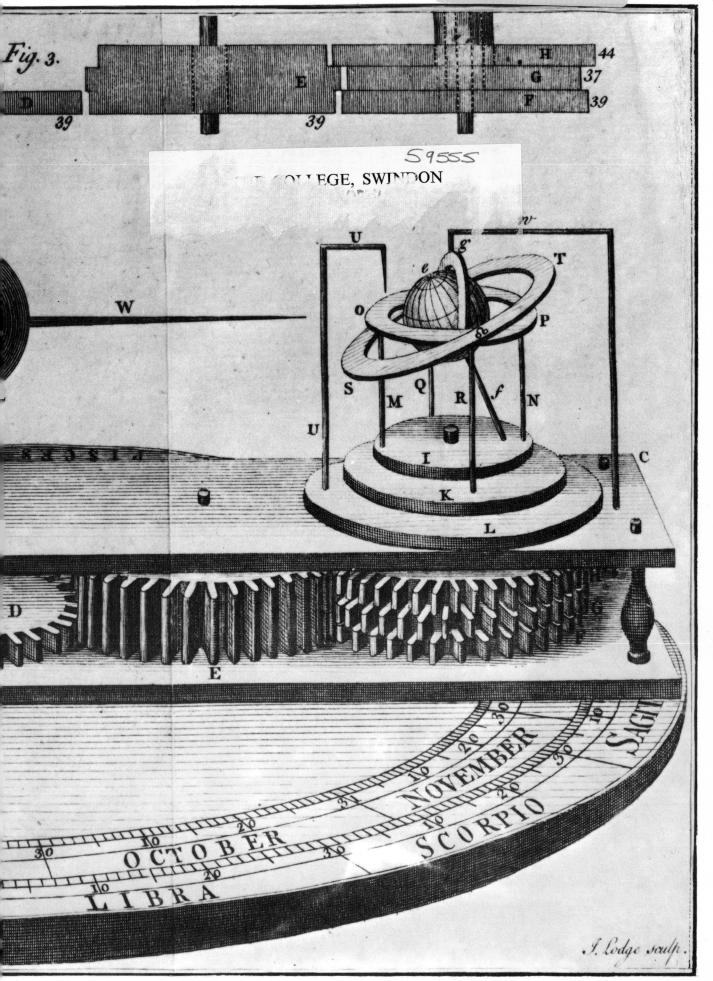

Fig. 3.

D
39

E

39

H   44
G   37
F   39

U

W

U

g
e
o
T
P
S   M   Q   R   f   N
I
K
L
C

D

E

6

LIBRA   OCTOBER   SCORPIO   NOVEMBER   SAGIT

J. Lodge sculp.

SCIENTIFIC INSTRUMENTS

Harriet Wynter and Anthony Turner

Studio Vista

'Man is a tool-using animal . . .
Without tools he is nothing, with
tools he is all'

Thomas Carlyle 1795–1881
Sartor Resartus book i, chapter 5.

Studio Vista
An imprint of Cassell & Collier Macmillan Publishers Ltd,
35 Red Lion Square, London WC1R 4SG,
and at Sydney, Auckland, Toronto, Johannesburg,
an affiliate of Macmillan Inc., New York

Copyright © Harriet Wynter and Anthony Turner 1975
First published in 1975
Designed by Flax & Kingsnorth

ISBN 0 289 70403 0

Set in Monophoto Baskerville
Printed by Jolly & Barber Ltd, Rugby, England

Frontispiece
**Instruments for a cosmographer,
engraving on paper 262 × 387 mm
($10\frac{1}{4}$ × $15\frac{1}{4}$ in.), between 1691 and 1701.**

**The ship *Argo* surmounts the globe in a
central laurel wreath. Above this the
winged figure of Fame blows a
trumpet on the banner of which is the
motto of the Argonauts, '*Plus Ultra*'.
Surrounding this is an elaborate
decorative ensemble of mathematical
instruments.**

Devised *c.* 1688.

**The second and more elaborate
emblem of the Accademia
cosmographica degli Argonauti
founded in 1684 by the globe-maker
and cartographer Vincenzo Maria
Coronelli (1650–1718). The design
seems to have been specially drawn
for use in Coronelli's folio works,
particularly the thirteen volume
*Atlante veneto nel quale si contiene la
descrittione . . . dell' universo . . .*,
Venice 1691–1701.**

CONTENTS

PREFACE

The purpose of this book is to provide an introductory description and guide to antique scientific instruments which, at the moment, are available to the collector. It seeks to provide a brief description of the main kinds of instruments employed in astronomy, navigation, making sundials, surveying and optics, together with an account of their development and uses. Of the earlier instruments, most have either disappeared or are safely in public collections, and few examples earlier than the sixteenth century are likely to be encountered. It is therefore with collectable instruments from the mid-sixteenth century to the mid-nineteenth century that the authors have been mainly concerned. Wherever possible the illustrations have been drawn from private sources and collections. Since this is so, it has seemed worthwhile to give as fully detailed descriptions of them as possible.

In writing the book, chapters 1, 2 and 4 have primarily been the work of Harriet Wynter, chapters 3, 5 and the captions of Anthony Turner. Naturally we have incurred many obligations and it is with much gratitude that we acknowledge the help we have received on numerous occasions from David Bryden, Dr Mary Holbrook, Francis Maddison, Reuben Shackman M.B.E., Alan Stimson, Inst Cdr A. G. Thoday, and Dr Helen Wallis. Lorna Graves, Maxime Melbourne and Jessica Rainsford-Hannay have immeasurably lightened the thankless task of typing the manuscript. We also offer our thanks to all the institutions and private collectors who have generously supplied photographs or allowed us to illustrate items from their collections. Finally we have to thank Jennifer Drake-Brockman for compiling the index.

Astronomy

1

ASTRONOMY is not only the oldest of the sciences, it is also the most important, for its principles are the basis of the rest. Primitive man had no other way for measuring time and noting the seasons than by the movements of the heavens. Over millennia, numerous forms of calendars, repositories of rudimentary astronomical information, were drawn up in primitive agrarian societies. However, it was only in the centres of urban information in the Middle and Near East, that such knowledge could be codified to supply the basis of a scientific, theoretical astronomy, and it was only at a very late stage in the development of civilization that this happened.

The development of astronomy, like that of the other sciences in Western Europe, rested upon the ideas and theories enunciated and refined in the classical Greek and Hellenistic worlds. The summit of Greek astronomical speculation, and its legacy to succeeding centuries, was Ptolemy's great systemization as set out in the *Syntaxis mathematicae*, better known as the *Almagest*. Ptolemy offered a geometrical model of the heavenly bodies by which their positions could be calculated and their behaviour predicted. He did not insist on its physical reality. Since it is the movements of the stars and planets in relation to man on earth which is the chief concern to the earthbound observer, it was perfectly reasonable to take the earth as the centre of the whole system, and this is what the Ptolemaic system does. The globe of the earth was considered as the centre of a vast celestial sphere against which the planets performed their complicated movements, and which could be divided into a grid against which they could be plotted. Since most astronomical instruments use this set of lines, it is to their description that we must next turn.

Definitions

As the earth hangs in space, two points set on its surface will be due north and south. These, the north and south poles, are diametrically opposite each other, and a straight line drawn between the two constitutes the earth's polar axis (see fig. 1). The equator is a line drawn on the circumference of the globe whose plane cuts the polar axis at right angles. Then consider the earth as hanging inside a larger sphere at its centre point (fig. 2). If the polar axis is extended to the larger sphere the points at which it cuts the surface will be the north and south celestial poles. Similarly, if the plane of the equator is extended to touch the celestial sphere, it provides the position of the celestial equator.

While the earth traces out its annual orbit around the sun, its motion appears to its inhabitants as the passage of the sun across their sky, and this apparent path of the sun is represented by a great circle in space termed the ecliptic.

Because it is that much closer, the moon appears to trace out a double loop around the earth, intersecting the ecliptic at four points, and when to the observer on earth, the moon and the sun are in a direct line of sight, an eclipse occurs.

As the year progresses, the sun rises with different groups of stars, termed constellations, which were given mythological characters by the ancients, and form the twelve signs of the zodiac. These twelve constellations lie along a belt which extends for 8° on either side of the ecliptic, and are loosely associated with the months of the year. All the planets except Pluto and a few asteroids wander through this band. Seven planets were counted in antiquity: the moon, Mercury, Venus, Mars, Jupiter, Saturn and the sun. The earth was not termed a planet until after Copernicus (1473–1543). Uranus was discovered by Herschel in 1781, Neptune in 1846 and Pluto in 1930.

The earth has an axis of rotation which is not perpendicular to its path around the sun, the angle between the two (the 'obliquity of the ecliptic') being 23° 27′. This angle changes slightly across the centuries, and archaeological evidence suggests it was 24° 22′ in 5000 B.C. During the solar year, the ecliptic twice intersects the celestial equator, and these points when the sun crosses the equator from south to north in the spring

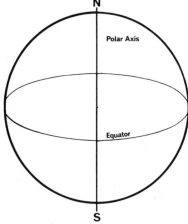

Fig 1
Diagram to show north and south poles, the earth's polar axis and the equator.

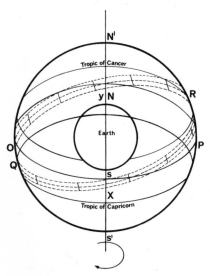

Fig 2
Diagram to show tropics, ecliptic, zodiac and equator according to the Ptolemaic system.

OXPY celestial equator
QXRY centre, dotted line, ecliptic
N¹S¹ celestial polar axis
X spring equinox Y autumn equinox
R winter solstice
Q summer solstice

Fig 3
Indo-Persian astrolobe, not signed,
nineteenth century, diameter 203 mm
(8 in.), brass, front and back views.

Rete for twenty-nine fixed stars with
curved star pointers and including
part of the equinoctial circle within
the ecliptic. Elaborately pierced *kursi*
with engraved floral decoration.
Simple shackle suspension. Five plates
of which the one visible beneath the
rete is engraved below the horizon line
with prayer lines and with dotted
lines for Babylonian hours.

Back shows in the upper left quadrant
a sexagesimal sine graph, and in the
upper right quadrant a diagram for
the meridian altitude of the sun. In the
lower quadrants is a shadow square.

9

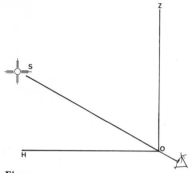

Fig 4
Zenith distance
If Z is the zenith point, O the observer
and S the celestial object observed,
then SZ is the zenith distance and
< SOZ its angular measurement, and
< SOH is the altitude or angular
elevation above the horizon.

Fig 5 *right*
Engraving of quadrants 1788, from
William Henry Hall and others *The
New Royal Encyclopedia; or complete
modern dictionary of arts and
sciences . . . 3 vols, London, 2nd ed.
(c.1789/90) vol. 1, plate 5.*

and the reverse in the autumn, and rises due east all over the world, are termed the equinoxes. At these equinoctial points, day and night are of equal length everywhere on earth, hence the etymology of the word, from the Latin *aequus* (equal) and *nox* (night).

Equidistant between the equinoxes are the solstitial points, which mark the maximum distance of the ecliptic from the celestial equator. In real terms the solstices occur when the sun is overhead at one of the tropics and our hemisphere is in midsummer or midwinter. Just as a pendulum is momentarily stationary at the end of its swing, the sun appears to hesitate, hence the term solstice which means 'the standing of the sun'. The tropics are two small circles on the celestial sphere which are parallel to the equator and pass through the solstices. The path of the sun lies between the tropics, the northern being the tropic of Cancer, the southern the tropic of Capricorn (see fig. 2).

In order to locate the position of stars on a celestial grid two angular distances are required. These are the declination (angular distance north and south of the celestial equator) and 'right ascension' of the star, which is the angular distance measured from the great circle through a hypothetical star known as the First Point of Aries, which is the point on the sky where the ecliptic crosses the equator in spring.

To obtain the declination of a star the meridian altitude is measured and its value added or subtracted from the observer's co-lat (i.e. the remainder left when the latitude is subtracted from 90°). This gives the star's declination.

The complement to altitude is zenith distance, an alternative sometimes found on astrolabes, which means the distance of a celestial object from the zenith (i.e. point vertically above an observer) measured in degrees (see fig. 4). The complement of declination is 'polar distance', a variable used on early instruments which has been discarded.

The declination and the right ascension have the same function as terrestrial latitude and longitude. While the earth is in orbit around the sun, it is also spinning like a top on its polar axis, and completes its revolution in near enough 24 hours. This is termed the diurnal motion, from the Latin *diurnalis* meaning daily, and supplies the fundamental division of time. There are, however, many different ways of dividing it up into hours (see chapter 3) and the point from which counting begins may also vary. The Hebrews used from sunset to sunset, 'the evening and the morning' while astronomers preferred from noon to noon, i.e. at the sun's meridian – from the Latin *meridianus* meaning midday or southern – when the sun reaches its highest point in the sky above the horizon each day, so that it did not interfere with their nocturnal calculations. Merchants from the time of Babylon onwards found midnight a more convenient division and the dispute was continued until 1925, when the astronomers finally surrendered.

Observational Instruments

Astronomical instruments can be divided into the categories instruments for observation and devices for calculation and demonstration. By the nature of the task which they had to perform, observational instruments could be made more accurate if they were larger and permanently fixed in position. Monumental observatory instruments, even where ancient ones survive, are beyond the scope of most collections and will not be considered here. Their design, however, affected that of other instruments. The new techniques pioneered in the making of these large instruments to the special orders of astronomers such as Tycho Brahe, Hevelius, Flamsteed and Halley, led to improvements in more everyday forms of instrument. In particular a steady sale of standard instruments provided the capital for the experiments required to refine accuracy.

The Quadrant
The original idea of a quadrant to take declination and zenith distances is attributed to Ptolemy, who advocated the use of a quarter of a circle rather

Plate 5

Various improved Quadrants, with Hadley's Sextant
See the Systems of Astronomy & Navigation

Fig.1

Fig.2

G. Adams London

Fig.1 Adams's small Quadrant.

Fig.2 Bird's twelve inch Quadrant.

Fig.3

G. Adams London

Fig.4

Fig.8

Fig.9

Fig.10

Fig.3 Mural Quadrant.

Fig.4 Hadley's Sextant.

Fig.5

Fig.6

Fig.7

Published as the Act directs by C. Cooke, No 17 Paternoster Row, Feb.ʸ 26ᵗʰ 1790.

Fig 6
English transit instrument, signed
'*Fait par Ramsden pour Dollond*',
c. 1780, length of telescope 1067 mm
(42 in.), brass.

A fine portable transit instrument,
with Y mounts, setting level and
illuminating annulus.

than a full circle because he assumed, rightly, that the arc could be graduated into more legible divisions. He designed a quadrant which used a plummet for vertical alignment, and, while there is no evidence that he built such an instrument, his plans were adopted with the rest of his writings. The principle of his quadrant required a plane enclosed by two radii, with an alidade (a moving sight arm) pointing through the apex, sighted from a graduated arc, which indicates the degrees of altitude or zenith distances of a star as required.

Monumental quadrants were used by astronomers until the end of the eighteenth century. When Edmond Halley (1656–1742) succeeded John Flamsteed at the Greenwich Observatory he found that Flamsteed's instruments had been removed by his relations as they asserted that he had paid for them. Fortunately Halley was given a grant of £500 with which he commissioned an 8-foot mural quadrant from George Graham for £73 together with a 5-foot telescope. The mural quadrant was fixed to a massive pier, where originally Flamsteed's sextant had been placed, and with these two instruments Halley never missed a meridian transit of the moon for over seventeen years. Graham divided the quadrant into two sets of graduations, and each scale checked on the other, so that readings could be made within 5 seconds of arc (fig. 5). 'Honest George Graham' was a member of the Royal Society and spent his life in scientific enquiry and the manufacture of suitable astronomical instruments for both Halley and his successor James Bradley, who wrote several years after Graham's death in 1751:

'I am sensible that, if my own endeavours have, in any respect been effectual to the advancement of Astronomy, it has principally been owing to the advice and assistance given me by our worthy member, Mr George Graham, whose great skill and judgement in mechanics, joined with a complete and practical knowledge of the uses of astronomical instruments enabled him to contrive and execute them in the most perfect manner.'

Mural quadrants were made by John Bird (1709–76) and Jeremiah Sisson (fl. 1736–88) for both English and foreign observatories, and the improvements and refinements they embodied reached a height of technical excellence which was disseminated throughout the instrument-making industry. Although the possibility of acquiring one of these large quadrants is remote, smaller instruments of 10–20 inch radius were made based on the same scientific principles of which several survive.

The commonest forms of observational quadrants, however, are those developed in portable form specifically for navigational or geodetic use, where smallness of size had to be reconciled with accuracy. Despite the distinction of function which is made it is worth emphasizing their common origin.

The invention of the telescope and the introduction of telescopic sights after 1670 led to a great increase in the accuracy which could be obtained with the quadrant, and also to production of new forms of instrument. Among these can sometimes be found the transit instrument used for measuring the exact moment at which a star or other pre-selected celestial body crossed a given meridian. This instrument most commonly occurs in small, portable versions intended for use in field survey work (fig. 6).

The Astrolabe

After the sack of Alexandria in A.D. 389 many scholars turned towards Syria and Persia. The concepts of Greek learning in science and technology were transmitted and absorbed by the races of the Middle East, a process which was only quickened after the incursion of Islam in the seventh century. Alexandria was sacked again in 641 and the great library dispersed by the followers of Mahommet. Under Mohammedan domination from 632, the Middle East provided the conquerors with the riches

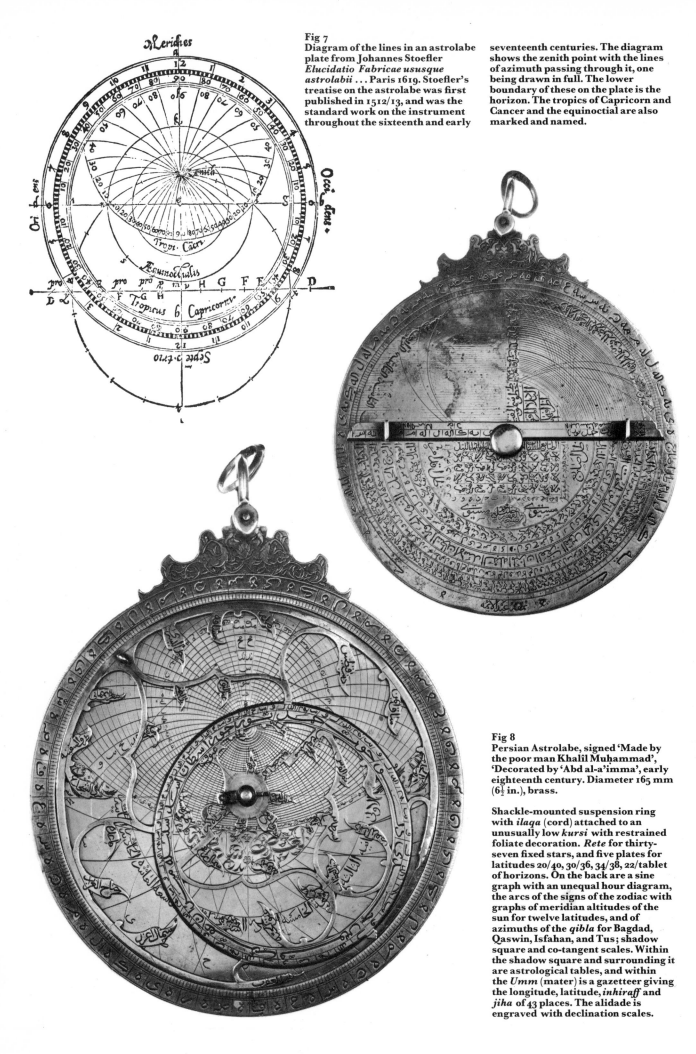

Fig 7
Diagram of the lines in an astrolabe plate from Johannes Stoefler *Elucidatio Fabricae ususque astrolabii* ... Paris 1619. Stoefler's treatise on the astrolabe was first published in 1512/13, and was the standard work on the instrument throughout the sixteenth and early seventeenth centuries. The diagram shows the zenith point with the lines of azimuth passing through it, one being drawn in full. The lower boundary of these on the plate is the horizon. The tropics of Capricorn and Cancer and the equinoctial are also marked and named.

Fig 8
Persian Astrolabe, signed 'Made by the poor man Khalīl Muḥammad', 'Decorated by 'Abd al-a'imma', early eighteenth century. Diameter 165 mm (6½ in.), brass.

Shackle-mounted suspension ring with *ilaqa* (cord) attached to an unusually low *kursi* with restrained foliate decoration. *Rete* for thirty-seven fixed stars, and five plates for latitudes 20/40, 30/36, 34/38, 22/tablet of horizons. On the back are a sine graph with an unequal hour diagram, the arcs of the signs of the zodiac with graphs of meridian altitudes of the sun for twelve latitudes, and of azimuths of the *qibla* for Bagdad, Qaswin, Isfahan, and Tus; shadow square and co-tangent scales. Within the shadow square and surrounding it are astrological tables, and within the *Umm* (mater) is a gazetteer giving the longitude, latitude, *inhiraff* and *jiha* of 43 places. The alidade is engraved with declination scales.

of learning, which they enthusiastically absorbed, and spread across southern France to Spain, to advance in navigation, mathematics, surveying and astronomy. A valuable tool in all these activities was the astrolabe.

Possibly known in theory to Hipparchus, and given the name 'astrolabium' by Ptolemy, the first known Arabic treatises on the astrolabe appeared in the early part of the ninth century and the earliest known surviving instruments were made in the tenth. Jewish scholars in Spain translated Arabic texts into Latin from the tenth century onwards, although few European astrolabes are known before 1200. The use of the astrolabe in the west declined from the end of the seventeenth century but it continued to be made and used in Islam until recent times. It has always been held in great esteem; to Chaucer it was 'a noble instrument' and to Abelard and Heloise a suitable name for their child.

The astrolabe is a two-dimensional map of the three-dimensional heavens. It is also an analog computer enabling its user to find answers to such problems as where a given star will be at a given time on a given date by a simple mechanical process, instead of by long and tedious arithmetical calculations. In addition blank spaces on the astrolabe back are used for various kinds of graphs and tables. The map is drawn on a stereographic projection, that is an observer is imagined to be looking at the celestial sphere from one of the celestial poles. From this point he

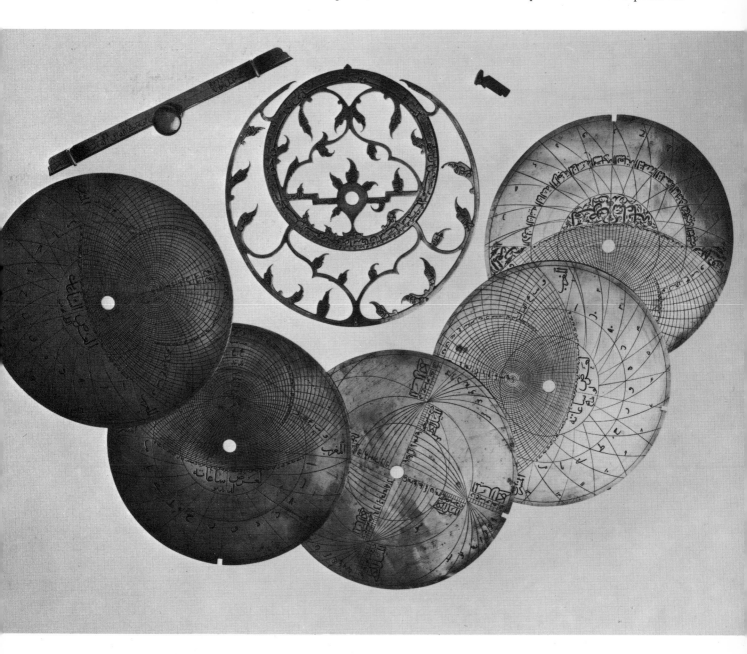

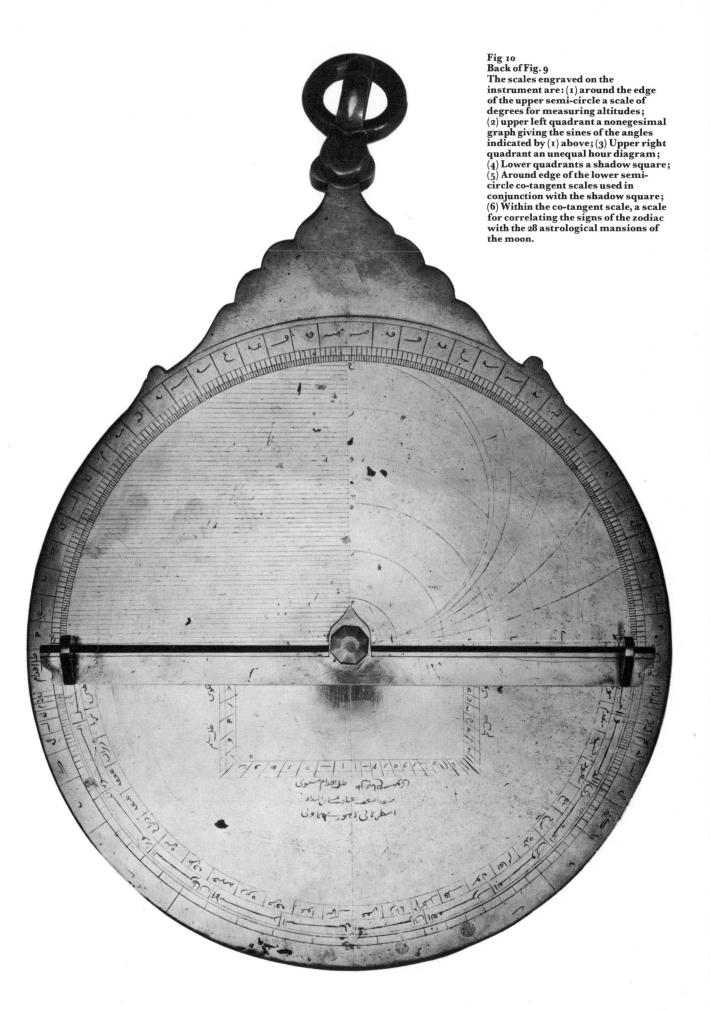

Fig 10
Back of Fig. 9
The scales engraved on the
instrument are: (1) around the edge
of the upper semi-circle a scale of
degrees for measuring altitudes;
(2) upper left quadrant a nonegesimal
graph giving the sines of the angles
indicated by (1) above; (3) Upper right
quadrant an unequal hour diagram;
(4) Lower quadrants a shadow square;
(5) Around edge of the lower semi-
circle co-tangent scales used in
conjunction with the shadow square;
(6) Within the co-tangent scale, a scale
for correlating the signs of the zodiac
with the 28 astrological mansions of
the moon.

projects onto the plane of the celestial equator all the celestial reference points (the tropics, ecliptic, etc.) and such stars as it is desired to plot. If the astrolabe is to be drawn for northern latitudes, then the centre of the projection is the south pole, if for southern, the north pole.

A valuable teaching instrument, it could be used for measuring angles, and could solve tedious calculations by mechanical means. It was used for many purposes, mainly for ascertaining the time, judging altitude, and for determining the position of celestial bodies at a particular moment in time. It was also used in mosques by the *muachad* for determining the astronomical time *(mu'addil)* when the faithful were called to prayer by the *muezzin*.

The astrolabe is made up of several parts, the Arabic alternative names are in brackets:

The MATER (Umm): This consists of a thick round plate, with a solid square-sectioned ridge termed a limb, and a shaped projecting shoulder *(kursi)* to take the suspension ring. On Islamic instruments the *kursi* is bold and decorated, for it symbolizes a throne with appropriate connotations, while on western instruments this is comparatively insignificant, with the notable exception of the Arsenius type which bears leaning caryatids.

The KURSI is aligned through the degree mark to the centre of the instrument, and bears a small shackle, which on Western instruments is usually mounted on a swivel joint, before it is connected to the suspension ring *(halka)*, but rarely so on Persian instruments.

Fig 11
Indo-Persian astrolabe, signed 'Made by 'Isà b. Ilah-dâd the royal astrolabist of Lahore'. The instrument disassembled showing the five plates for latitudes 18°/29°; 27°/32°; 37°/40°; 0°/combined projection for 24° and 30°; tablet of horizons/tablet of ecliptic co-ordinates. The interior of the mater is engraved with a gazetteer of the latitudes and longitudes of various places, arranged by regions.

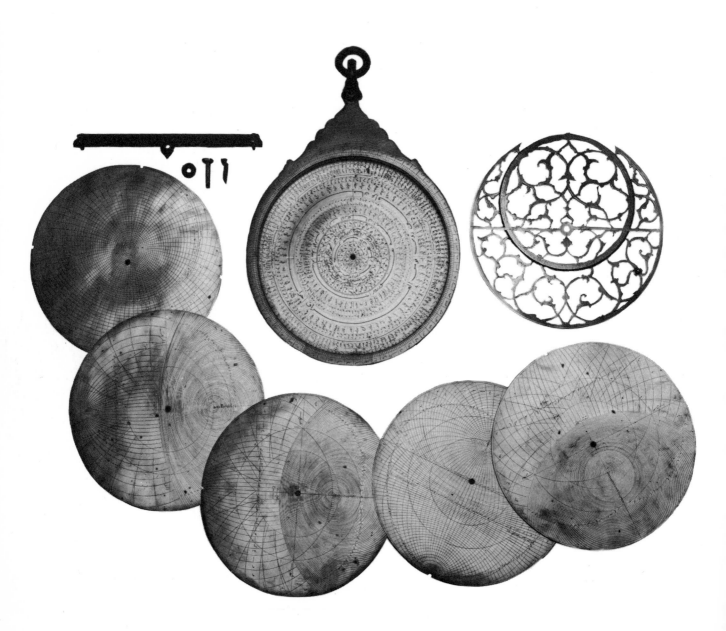

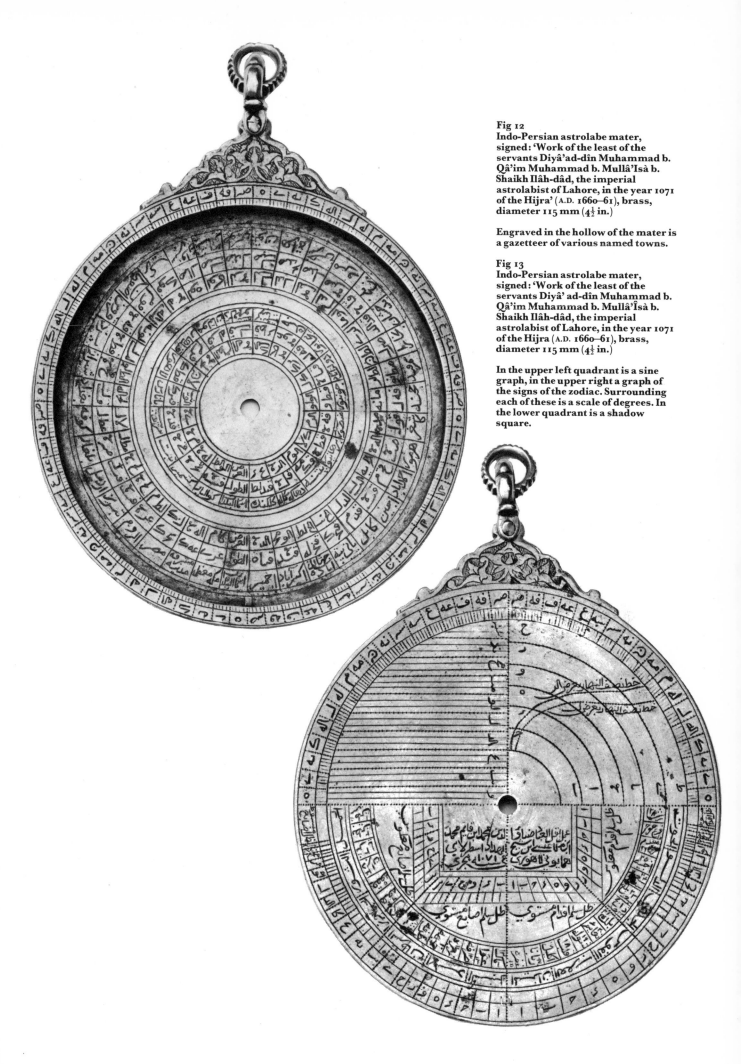

Fig 12
Indo-Persian astrolabe mater, signed: 'Work of the least of the servants Diyâ'ad-dîn Muhammad b. Qâ'im Muhammad b. Mullâ'Isà b. Shaikh Ilâh-dâd, the imperial astrolabist of Lahore, in the year 1071 of the Hijra' (A.D. 1660–61), brass, diameter 115 mm (4½ in.)

Engraved in the hollow of the mater is a gazetteer of various named towns.

Fig 13
Indo-Persian astrolabe mater, signed: 'Work of the least of the servants Diyâ' ad-dîn Muhammad b. Qâ'im Muhammad b. Mullâ'Îsà b. Shaikh Ilâh-dâd, the imperial astrolabist of Lahore, in the year 1071 of the Hijra (A.D. 1660–61), brass, diameter 115 mm (4½ in.)

In the upper left quadrant is a sine graph, in the upper right a graph of the signs of the zodiac. Surrounding each of these is a scale of degrees. In the lower quadrant is a shadow square.

19

The centre of the mater is pierced to take a round-head pin with a slit in its side, through which a wedge termed a horse *(faras)* can be passed to hold together all the movable parts inside the mater. The plates or tympans *(safiha)*, discs of brass which vary in number, slot into the limb in the centre of the *kursi*, and more rarely, are held by a central protrusion into a hollow in the mater.

A skeletal brass disc termed *rete* in Latin, meaning net *(Ankabut)* is placed over the plates, occasionally with a rotatable rule on European examples. The back of the mater bears a sighting arm, a device with two aligned pinnules *(alidade)*. To assemble an astrolabe, one inserts the pin into the alidade, through the back of the mater, the plates, and the *rete*, and sometimes a washer *(fals)*, then the horse locks with the pin, and everything is held in place.

The scales and divisions on the instrument vary. On Islamic astrolabes the degree scale of 360° is engraved on the limb, whereas on western examples it can be 24 equal hours in 1–12 twice, with or without the 360°. Sometimes the western instrument bears a quadrantal scale, i.e. 0°–90° repeated four times. The 0 point is usually under the *kursi* but it can also be placed at 9 o'clock.

The plates are engraved, usually on both sides, for different latitudes with a number of circles and intersecting arcs, which are seen to be concentrated around a central point immediately below the degree mark on the *kursi*. This central point represents the observer's meridian, and the centre of the complex of arcs, his zenith. The radiating arcs from the zenith are *azimuths*, or compass bearings, and the circles intersecting them are *almacantars*, the circles of equal altitude measured in degrees. Through the centre of the plate is the east-west line, and the last almacantar represents the horizon. Eighteen degrees below it another arc represents the twilight or crepuscular line which marks the moment before dawn and after sunset before day or night has begun. Between the two is the latitude of the observer's geographical position. As we are presumed to be surveying the imaginary grid from the celestial south pole, the outer circle running round the circumference of the plate represents the tropic of Capricorn, the next circle inside it, the equator and the third, the tropic of Cancer, and radiating from this are twelve arcs which represent the twelve unequal hours, i.e. the twelve divisions each of day and night, which are uneven in length except at the equinoxes (fig. 7).

The RETE: the skeletal disc which covers the appropriate place, sometimes compared with a spider's web, is a star map. Bounded like the plates with the circle of the tropic of Capricorn, the inner offset circle on the rete *(ankabut)* represents the band of the ecliptic, on which the constellations of the zodiac are to be found. The beautifully engraved leaves are pointers to a variable number of bright stars depending on the size and complexity of the instrument.

The back of the mater varies considerably. On European instruments the scales can be read with ease. Starting from the outer edge these are the degree scales, zodiac scale and calendar scale (Julian or Gregorian; first point of Aries should be 21 March for Gregorian). In the upper right-hand quadrant are the arcs of the unequal hours (see chapter 3), and below the central east-west line there is a rectangle, twice as long as it is high, termed the shadow square, which is used for altitude measurement. The horizontal side is termed the *umbra recta* and the vertical side is termed the *umbra versa*, as they were first used for taking the height of the sun by its shadow. A knowledge of Arabic is useful for Islamic instruments.

In the upper left-hand quadrant there is frequently a sine graph for assisting trigonometrical calculations, and graphs of meridian altitudes of the sun for a number of latitudes and azimuths of the direction of Mecca *(qibla)* from other cities. Some astrolabes have co-tangent tables and astrological tables. On the inside of the mater itself there is often a gazeteer of longitude and latitude *(inhiraff)* and *(jiha)* of different places. On the Persian astrolabe in fig. 8 there is a cartouche bearing an inscription of pride of workmanship in the oriental manner of self effacement.

Fig 14
Maghribî astrolabe, not signed, early
nineteenth century, brass.

Rete with twenty-three fixed stars,
four *mudirs* (knobs for rotating the
rete), placed one in each quarter.
Plain *kursi* with restrained scroll
work, suspension apparatus with
shackle and swivel ring. On the back
is a shadow square, degree scale and
zodiac/calendar scale (concentric
type).

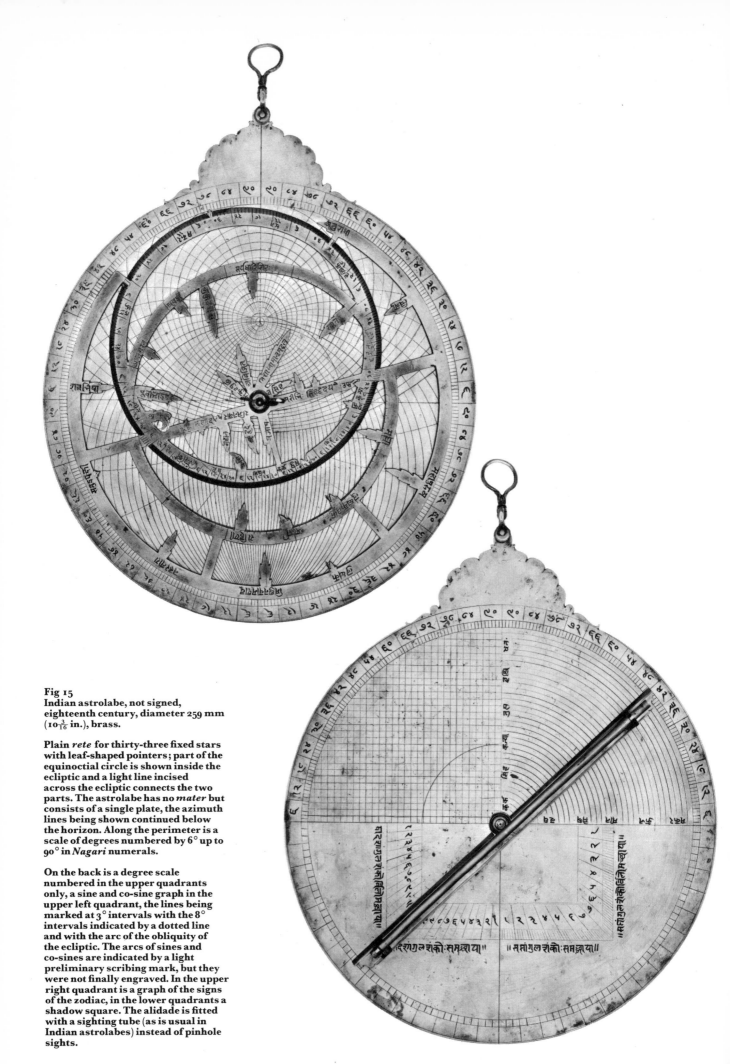

Fig 15
Indian astrolabe, not signed,
eighteenth century, diameter 259 mm
(10$\frac{3}{16}$ in.), brass.

Plain *rete* for thirty-three fixed stars
with leaf-shaped pointers; part of the
equinoctial circle is shown inside the
ecliptic and a light line incised
across the ecliptic connects the two
parts. The astrolabe has no *mater* but
consists of a single plate, the azimuth
lines being shown continued below
the horizon. Along the perimeter is a
scale of degrees numbered by 6° up to
90° in *Nagari* numerals.

On the back is a degree scale
numbered in the upper quadrants
only, a sine and co-sine graph in the
upper left quadrant, the lines being
marked at 3° intervals with the 8°
intervals indicated by a dotted line
and with the arc of the obliquity of
the ecliptic. The arcs of sines and
co-sines are indicated by a light
preliminary scribing mark, but they
were not finally engraved. In the upper
right quadrant is a graph of the signs
of the zodiac, in the lower quadrants a
shadow square. The alidade is fitted
with a sighting tube (as is usual in
Indian astrolabes) instead of pinhole
sights.

When using it in sunlight an ancient astronomer held the astrolabe at waist level by its cord *(ilaqa)* attached to the suspension ring, and turned it into the plane of the sun. He then adjusted the alidade until the sunlight passed through the two pinnules, and took a reading from the scale so that the sun's height from the horizon, its almacantar, could be determined. Using it at night, he would hold the instrument at eye level so that a given star could be sighted through the two pinnules. With this information, he could set the rete. Having selected the appropriate plate for his latitude the observer turned the rete over it until the observed celestial body was on the almacantar measured with the alidade. The instrument was then an exact reproduction of the heavens at that moment, with all the celestial bodies arranged correctly in relation to each other. If desired the instrument could be set to represent the pattern of the heavens at any period of time, past, present, or future.

A disadvantage of the astrolabe was that its use was limited to the number of plates available, each for a separate latitude. Even if cost and bulk were disregarded, the instrument could never be comprehensive enough for travellers. 'Universal' astrolabes, with one plate only to represent a whole hemisphere, were developed to overcome this disadvantage.

Three types were developed, in the sixteenth and seventeenth centuries. The most widely used was based on the calculation of eleventh-century Toledan scholars, the *saphaea arzachelis*, which was revived by instrument-designer Gemma Frisius of Louvain as the *astrolabum catholicum* and took as its point of reference the vernal equinox instead of the celestial pole.

An account of a universal astrolabe written by Hugo Helt, a Frisian attached to his father's household, was published in Paris by Juan de Rojas of Sarmiento, which has since been known as the Rojas projection. This type of astrolabe does not use stereographic projection, but orthographic projection in which the centre of projection is considered to be a point at infinity, perpendicular to the plane of the solstices, which remains the plane of projection as in the *saphaea arzachelis*. It differs from the Gemma Frisius universal mainly in that the meridians familiar as arcs of circles are replaced by semi-ellipses (fig. 31).

The third type of universal was developed at the end of the seventeenth century by Philippe de la Hire who took as the centre of projection a point considered to be between those used on the preceding types, such that the meridians could be drawn as arcs of ellipses, and, being nearly equidistant, were easier to read.

No collection of astronomical instruments is complete without at least one astrolabe, but they are becoming increasingly difficult to find and can command high prices. There are a large number of fakes or copies, which are fun but cannot be considered seriously.

Gunter quadrant

A development from the astrolabe was the so-called Gunter quadrant. Edmund Gunter (1581–1626) invented a small pocket-size quadrant in about 1618, which he first described in 1623. It was adapted from the hand quadrant, and bore a number of scales derived from an astrolabe. Gunter's quadrant was a simple mechanical device for making astronomical calculations and computations, with a stereographic projection of the equator, ecliptic, horizon and tropic of Cancer designed with the eye at the north pole and the scales for date and time finding calibrated for the latitude of London.

The same principles apply for using a Gunter as for any other kind of quadrant (see pp. 10–11). When a celestial body is aligned through the peep sights on the radius, a plumb bob falls across an arcuate scale on the limb to indicate the astronomical information required.

The popularity of the Gunter quadrant has led to the manufacture of fakes. While several genuine versions of the Gunter exist like that with the Rojas projection (fig. 22) and that made at Oxford by John Prujean (fig. 19) a danger lies in those which sometimes appear to be faultless in

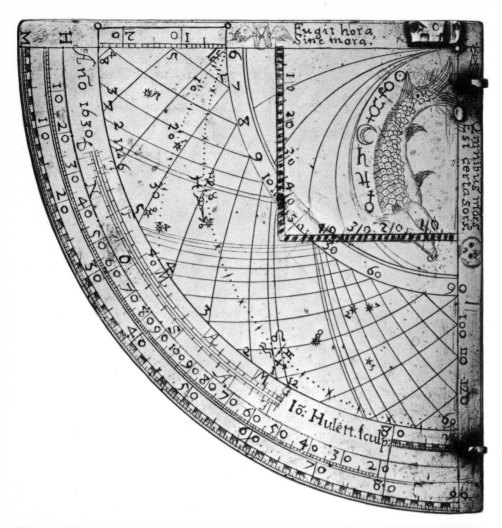

Fig 16
Face of an English quadrant, signed
'Anō 1630 Iõ: Hulett, sculp:' radius
144 mm. (5¾ in.), brass.

Along the limb is a 0–90° scale used
for making altitude observations in
conjunction with the pinhole sights
attached to one edge and a plumb line
(missing). Within this is a calendar
scale for declination and a Gunter-
type horary quadrant. The positions
of seven stars are marked on the
projection. In the apex is engraved the
figure of a monstrous fish
surrounded by planetary symbols.
Surrounding this is an unequal hour
diagram. A quadrant scale
(corresponding to the shadow square
of an astrolabe) which is engraved
over the fish and the unequal hour arcs
may be a latter addition. A winged
sandglass and a death's head are
engraved in the margins of the
instrument, accompanied
respectively by the legends, *Fugit
hora; Sine mora* and *Omnibus mors
est certa sors* ('Time flies without
pause' and 'Death is a sure fate for
all').

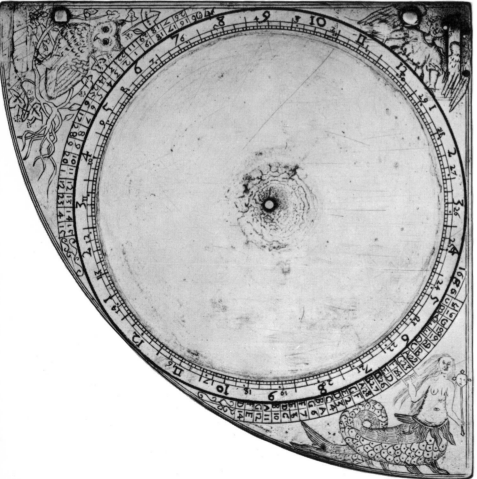

Fig 17
Back of an English Gunter quadrant
signed 'Anō 1630 Iõ: Hulett, sculp:'
radius 144 mm. (4⅛ in.). Brass.

The circular hour-scale (1–12 x2) and
date scale were for use as a nocturnal
in conjunction with a rotatable disc
now missing. The corners of the
instrument are embellished with
lightly engraved figures of an owl,
angel and mermaid with a fool's head.
Between the hour scale and the figures,
tables of dates and dominical letters
have been added by a later hand (or
hands), the dates '1686' and '1706'
being engraved at the head of each
series.

John Hulett (1607–1663) taught
mathematics in Oxford after
travelling widely throughout Europe.
He devised a projection for pillar dials
and wrote a lucid short account of the
Gunter quadrant, which was
published in 1665.

execution but are useless to operate. I have seen one with the projection copied from an astrolabe tympan. A peculiar similarity in their solid construction, style of cross-hatching and script in about half a dozen or so which have turned up since 1972, ostensibly from family collections. Whether these were made in the nineteenth century, however, or later, is not at all clear. The intention behind these instruments may not have been to deceive, but too little is known of their origin to enlarge on their history. Fortunately, sufficient is known about the instrument to prevent these copies being included in a serious collection, like the unfortunate beneficiaries of fake astrolabes.

The Gunter quadrant was a favourite instrument to be made by the amateur, such as the example made by the Oxford scholar John Hulett in 1630 (figs. 16 and 17) soon after the publication of Gunter's paper, while quadrants made by amateurs perhaps after a published illustration are not unknown (fig. 18). Learned books instructed the student on how to make his own instruments sometimes including a printed cut-out to be pasted onto cardboard, so that a cheap calculating instrument could be acquired. Gunter quadrants were also made as part of a compendium like the outer case of the example probably by Henry Sutton (fig. 20).

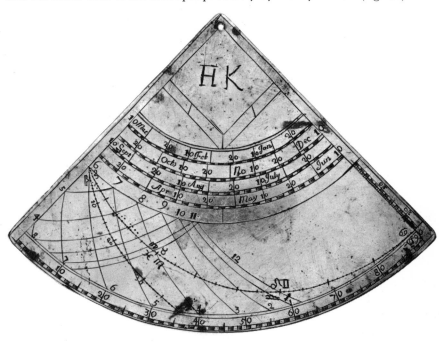

Fig 18
English quadrant, c. 1677. Radius 85 mm (3⅜ in.). Not signed but with the initials 'H K' in the shadow square, brass.

On the face is a Gunter-type quadrant, with 0–90° scale on the limb, used for making altitude observations in conjunction with the pinhole sights attached to one edge and a plumb line now missing. In the apex is a shadow square.

The back of the instrument is engraved with a perpetual calendar, a table of dominical letters from 1677 to 1704, a table of epacts for the date of Easter, and a coat of arms which may be those of its maker or owner.

The construction of the popular Gunter quadrant supplied valuable instruction in practical mathematics and examples made by gentleman amateurs, such as the present one, are sometimes found.

Fig 19
English quadrant signed 'Johā Prujean Fecit Oxon', *c.* 1680.

Length of side (this is not a radius) 106 mm (4 3/16 in.), brass.

On the face is a Gunter quadrant with six named stars, a shadow square, zodiacal calendar and 0–90° scale on the limb. At the apex is a hole for a plumb line (missing) used in conjunction with the sighting pinnules mounted along one edge. On the back is a rotatable disc engraved with an orthographic projection of the hour circles and parallels of declination onto the plane of the meridian. The positions of thirty-six stars are indicated and parts of two constellation figures. On the circumference of the disc are engraved a scale of degrees in four quadrants and a month scale. Surrounding the disc on the main body of the quadrant is a 24-hour scale (1–12x2). Mounted at the centre is a rotatable and graduated cursor.

The form of sundial which employs an orthographic projection is sometimes referred to as a 'Geminus' dial, but is more commonly known as a Rojas dial from its similarity with the Rojas astrolabe projection. The projection was used on a number of sixteenth-century instruments, but had been newly described by Richard Holland, mathematical tutor of Hart Hall, Oxford, in 1676. Prujean was closely associated with Holland and it was to accompany his work and following his design that he made quadrants of the type shown here. See Richard Holland *An Explanation of Mr Gunter's Quadrant. As it is enlarged with an Analemma* Oxford 1676.

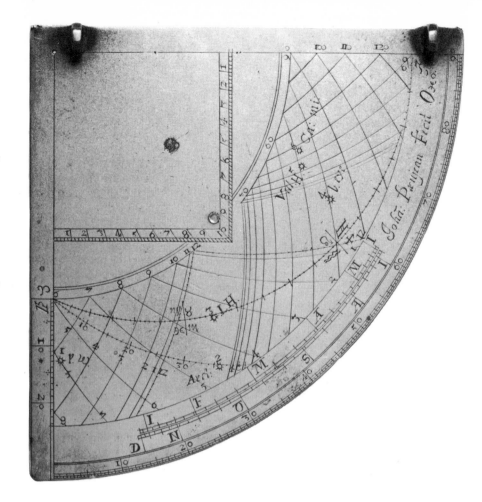

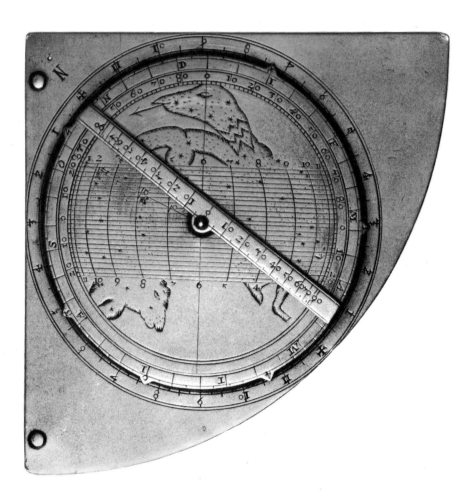

Fig 20
Face of quadrant not signed, *c.* 1700,
length of side (this is not a radius)
177 mm (7 in.), brass. A Gunter
quadrant with calendar and degree
scale (0–90°) reading to halves on the
limb. Within the shadow square is an
unequal hour diagram *(quadrans
vetus)*. The hole at the apex is for
attachment of a plumb line used in
conjunction with the sighting
pinnules set along one side.

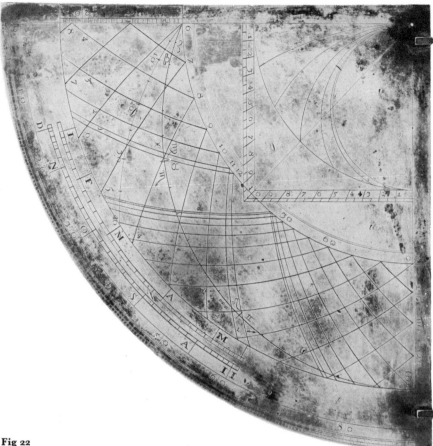

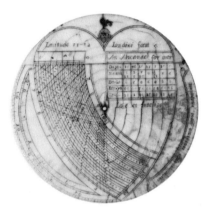

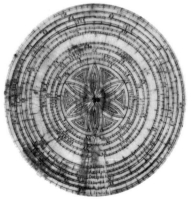

Fig 21
English Gunter quadrant, not signed,
but probably by John Browne. *c.* 1675,
ivory, diameter 75 mm (3 in.).

Circular disc engraved on one side
with a Gunter quadrant marked
latitude 51° 32′, and with a perpetual
calendar with the exhortation
lege et intelige (sic) 'read and
understand'. Both the calendar and
the quadrant are of the kind described
by John Browne and used on
instruments signed by him (*cf.* fig.
70). On one side of the lower edge is a
scale of degrees and a shadow scale.
On the opposite section is the
inscription *Perpetuus Iolis destinguet
tempora motus*. On the back is a
calendar scale surrounding a central
geometrical design, with
corresponding scales for the rising,
amplitude, declination, position in the
zodiac and right ascension of the sun
and a scale for the length of the day.

Fig 22
Back of quadrant, *c.* 1700. In the centre
is an orthographic projection of the
sphere similar to that used in a Rojas
universal astrolabe. This is
surrounded by a degree scale in four
quadrants with its zero point on the
East-West axis. On the limb is a
degree scale (0–90°) reading to halves.

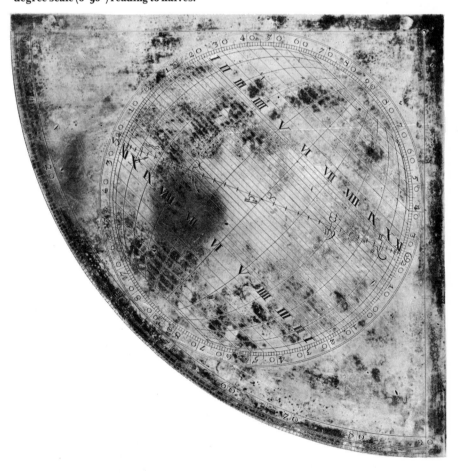

Spheres

Celestial spheres or globes represent stars as seen from outside, so they are in fact reversed from the observed positions inside the sphere of the heavens. They can be pivoted to represent the daily rotation of the stars and the axis can be tilted to correspond with the latitude of the geographical position. The globe also has a horizon ring at right angles to the axis to show which constellations are above and below the heavens at a given time, thus it can be used in determining the time of the rising or setting of the sun or a star.

Celestial globes were made in Islam throughout the Middle Ages and up to the eighteenth century. Made of brass, by the *cire perdu* process, they were engraved with star positions and the various celestial circles. Few globes seem to have been made in the medieval West, but by the late fifteenth century a new method was adopted, using shaped gores mathematically designed to fit over and completely cover a sphere made of lath and plaster. At first they were hand-drawn on parchment but were later printed on paper (the first printed map was published in 1472) and mounted into wooden stands. They gradually increased in size, until they became considerable items of furniture for the library or study.

It was not until the beginning of the sixteenth century that the terrestrial globe was produced as a companion piece.

The popularity of large globes was assured from the very start and they have continued to be made without a break until the present day (figs. 23, 24, 25 and 26). But by 1900 the globes that were made were fit

Fig 24
**Pair of Dutch globes, signed
'Guillielmus Jansonius Blaeu Auctor',
c. 1690, overall diameter of horizon
ring 477 mm (18¾ in.), printed paper
gores coloured by hand, brass
meridian ring, wood stand.**

**12-in. globes in original turned stands.
The terrestrial globe is dedicated to
Jacobus de la Feuille, the celestial to
Prince Maurice of Nassau.**

**William Janszoon Blaeu (1571–1638)
produced some of the most
outstanding maps and globes of the
seventeenth century. Having spent
some time studying with Tycho Brahe,
Blaeu employed his results for
positioning the stars in his celestial
globe, which was further
distinguished by the five pictorial
constellation images. Editions of the
globes were produced throughout the
century.**

only for the schoolroom, for, while accurate and clear, they were without
adornments and cannot be compared to their illustrious ancestors. As
they became out of date, old globes were allowed to deteriorate in attics
and stables. Restoration being extremely difficult, with very few craftsmen available capable of a proper job, it is with considerable sadness
that these poor relics are greeted when they are put up for sale because of
the high prices they now command.

Armillary sphere

Just as valuable a teaching instrument as the celestial globe was the armillary sphere. It consisted of a skeletal globe made up of the imaginary
circles, the equator, the tropics, arctic and antarctic circles, the meridians
passing through the poles at the first point of Aries and Libra to indicate
the vernal and autumnal equinoxes, and the oblique band of the ecliptic
divided into the twelve signs of the zodiac. A small sphere in the centre of
this grid represented the earth, with radial arms the moon, sun and sometimes the planets around it. The whole skeletal frame slotted into a horizon
ring mounted on an elaborate stand formed of four quadrants with a
baluster support and turned feet often beautifully chased and decorated.
With the most sophisticated of these, problems in connection with the
times of azimuths of sunrise and sunset could be solved. However, it was
never really a satisfactory calculating instrument, but aesthetics justified
its continuation as a demonstration model for teaching astronomy.

Fig 25

English celestial globe signed 'Manufactured and sold Wholesale and Retail by W. & T. M. Bardin, 16 Salisbury Square, Fleet Street, London' c. 1810–20. Diameter 460 mm (18 in.).

Globe with central core covered in paper gores printed from a copper plate hand-coloured and varnished. Brass meridian ring and fittings. Mahogany stand with turned central column and three cabriole-type feet on runners. The stretchers support a circular glazed box containing a compass needle and printed paper with 36 named points.

The firm of Bardin was recognized in its day as one of the leading globe-makers. The 'New British Globes' as they were called, were manufactured under the direction of W. & S. Jones, William Jones having himself computed the celestial globe. According to Thomas Keith a pair of the globes were placed in the Royal Observatory, Greenwich, by Maskelyne, to whom the celestial was dedicated.

Fig 26 *right*

English celestial globe signed 'NEWTON'S NEW & IMPROVED CELESTIAL GLOBE on which all the Stars, Nebulae & Clusters contained in the extensive Catalogue of the late E. Wollaston F.R.S. are accurately laid down, their Right Ascensions and Declinations having been recalculated for the Year 1840, by W. Newton.

'Manufactured by Newton & Sons, Chancery Lane, London Published January 1st 1845'. Overall diameter of horizon ring, 438 mm (17¼ in.).

Beechwood, brass and printed paper, with brass fittings; 12-inch globe of twenty-four gores with prominent stars. The constellation figures are depicted in dotted outline and lightly coloured in pastel shades. The globe is thus intermediary in the development from the elaborately pictorial celestial globes of the seventeenth and eighteenth centuries (see fig. 24) to the more austere examples later in the century which repudiated all constellation images (see fig. 29).

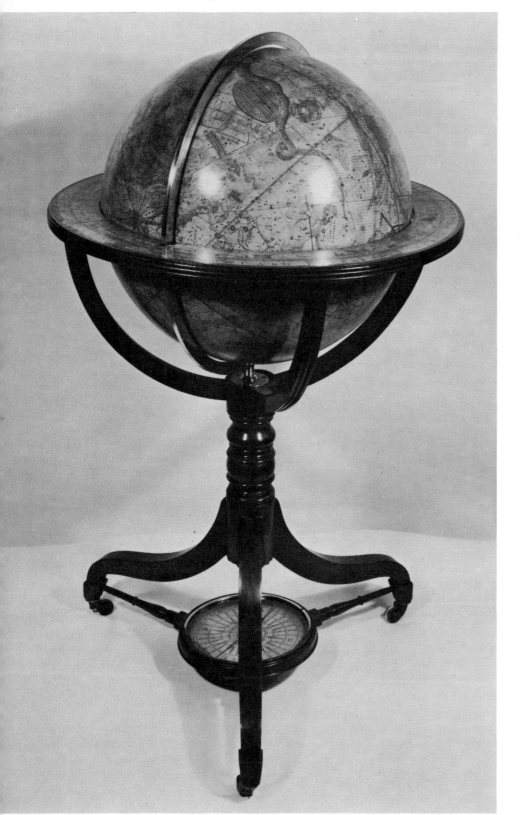

Fig 27
Miniature armillary sphere, not signed, sixteenth century. Diameter 55 mm (2⅛ in.). Gilt brass, with traces of vermilion colouring in the stamping.

The rings for the arctic and antarctic circles, the equator and tropics of Cancer and Capricorn are named in Latin as are the solstitial and equinoctial colures. The ecliptic/zodiac circle is divided on both sides in degrees for each sign.

The sphere is mounted on a later brass stand and was probably removed from an astronomical clock or other large instrument.

Fig 29 *below*
English globe, signed 'Smith's Celestial globe, showing the Number; Order and Magnitude of upwards of 4000 stars, on which are laid down all the stars contained in the catalogues of the most celebrated astronomers', 'Smith & Son', *c.* 1870, diameter 248 mm (9¾ in.).

Printed paper gores on *papier mâché* and plaster core with mahogany base and brass meridian ring. Only the zodiacal figures are shown pictorially on the globe which is mounted in a semi-circle meridian ring on a turned pillar with flat triform base.

Fig 28
Title page of Ioannis De Sacrobusto *Libellus De Sphaera. Accessit Eiusdem Avtoris Computus Ecclesiasticus, Et alia quaedem in studiosorum gratiam edita.* CVM PRAEFATIONE *Philippi Melanthonis. Impressum Vitebergae apud Vitum Creutzer. Anno,* M.D. XLV.

John of Holywood, Little book of the sphere, to which is added the ecclesiastical reckoner of the same author, and certain other works set forth for the benefit of students, with a preface by Philip Melancthon.

Foolscap octavo, coloured device of sphere on title page, woodcuts and diagrams in text, two with movable volvelles, and one with movable pointer, two folding tables. Contemporary MS annotations (mainly in the earlier part of the text). Ownership inscriptions of Leonhard Baldussius, Halle 1568; Andreas Raselius, 18 July 1578; Cornelius of Regensburg, 1632; Bookplate of the Royal Meteorological Society, Symons bequest, 1900.

Nineteenth-century half vellum and marbled boards.

Ioannes de Sacrobosco (*fl.c.* 1230). His treatise of the sphere has been described as the 'clearest, most elementary, and most used textbook in astronomy and cosmography from the thirteenth to the seventeenth century'. Combining traditional western astronomical knowledge, which ultimately stemmed from Macrobius and Ptolemy, with the fresh ideas derived from the twelfth century Latin translations of Arabic writers, particularly Alfraganus, the work perhaps dates from early in the thirteenth century, before 1220.

Manuscript copies of the treatise on the sphere were numerous in the Middle Ages, and it was translated into Hebrew by the Provençal scholar Solomon Abigdar in 1399. The advent of printing redoubled the number of available copies.

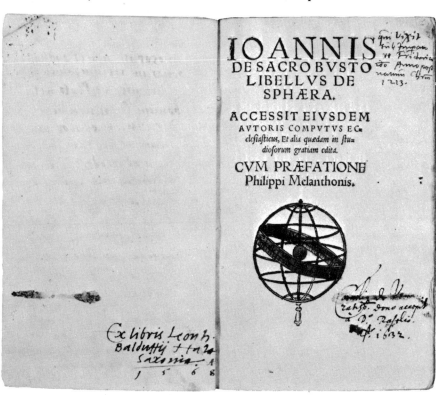

Resplendent brass armillaries of the sixteenth century are extremely rare, so while the dainty gilt brass miniature in fig. 27, probably from the top of an astronomical clock, cannot vie with the splendour of those in public collections, it can at least glory in sharing their date of birth. While brass armillaries were made throughout the seventeenth and eighteenth centuries, many older ones were repaired and new stands made. A rare eighteenth century example by Thomas Heath is illustrated in fig. 30.

Invented in eighteenth-century France, *papier mâché* instruments were made from about 1760. These charming toys were made in pairs to show the differences between the Ptolomaic and the Copernican systems, and were often displayed in schoolroom and library. The pretty instrument in fig. 32 is a delightful example, made by S. Fortin, rue de la Harpe, Paris, a maker of whom very little is known. Another maker who appears to have

Fig 31
Rojas type universal astrolabe, late seventeenth or early eighteenth century.

Fig 30
English armillary sphere signed 'Tho; Heath Fecit' *c.* 1730. Diameter of horizon ring 218 mm (8½ in.), overall height 277 mm (10⅞ in.). Brass, with ivory sphere representing the earth.

Removable sphere composed of arctic and antarctic circles, tropics of cancer and capricorn, equator and ecliptic/zodiac, held in a contemporary meridian ring which appears to have been originally intended for another instrument. The earth is mounted on a rotatable brass rod, mounted at the ecliptic pole. The whole sphere is contained in a horizon ring engraved with a zodiacal calendar, mounted on a turned single stem upon which it is rotatable, on an inverted domed base with lead weighting.

Fig 32 *left*
French Ptolemaic armillary sphere,
1780–1800; overall height approx.
403 mm (15⅞ in.). Beechwood, brass
and pasteboard, the rings covered
with engraved and varnished paper;
their circumferences painted in red.

Turned single stem stand on circular
base carrying horizon ring with four
supports engraved with the latitudes
and longitudes of various places.
Horizon ring with zodiacal calendar
having pictorial representations of
the signs of the zodiac. Into this slots
the meridian ring carrying a four
quadrant degree scale, and with a
24-hour dial at the north pole.
Mounted within the meridian circle
and free to rotate is a sphere formed
of armillary rings (named) for the
arctic, antarctic, the tropics of
Cancer and Capricorn, the equator
and the ecliptic/zodiac. The solstitial
and equinoctial colures carry degree
scales. On a spike in the centre of the
rings is a terrestrial globe.
Suspended from the ecliptic axis are
strips carrying sun and moon
emblems, rotatable in the plain of the
ecliptic.

Fig 33
French Copernican armillary sphere,
c. 1800, overall height approx. 410 mm
(16⅛ in.). Beechwood, brass and
pasteboard, the rings covered with
engraved and varnished paper,
coloured red, green and yellow.

Turned single stem stand carrying a
sphere composed of rings for the
solstitial and equinoctial colures, and
the ecliptic/zodiac band on which the
months and various signs are named
in Italian. Central gilt globe for sun
around which rotate four concentric
rings mounted on the central axis and
representing earth, Mars, Jupiter,
and Saturn. The earth on an inclined
axis is shown by a globe with
meridian ring and hanging moon
emblem. The proper periods of each
planet shown are marked on the rings:
Earth, 365 days, 5 hours, 49 minutes;
Mars, 1 year, 321 days, 22 hours, 19
minutes; Jupiter, 11 years, 315 days,
14 hours, 12 minutes; Saturn, 29 years,
163 days, 14 hours, 8 minutes.

made this particular form of instrument was the globe-maker C. F. Delamarche who was working into the middle of the nineteenth century. Some unsigned examples have survived, which resemble examples of the two known makers very closely, and may have been made by them for retailers (fig. 33). The Copernican part of the pair is awkward and ill-suited to the armillary form. The ecliptic is fixed horizontally enclosing moving concentric rings marked with the names of the planets whose orbits they represent. As the rings get smaller, the utter impossibility of portraying the tellurium (the earth and the moon) system in this manner becomes apparent. As these *papier mâché* instruments were made for foreign markets, the ecliptic was frequently printed in Italian or Spanish while the rest remained in native French, in an attempt to satisfy the customer.

Fig 34
Copernican armillary sphere, c. 1800, overall height 569 mm (22½ in.); diameter of horizon ring 385 mm (15¼ in.). Wood, brass and iron. The tôle or pontypool process has been employed in the construction of the rings.

Removable sphere with rings for the arctic circle, tropic of Cancer, equator, ecliptic and zodiac band, tropic of Capricorn, and antarctic circle. Each ring has a paper covering bearing its name. The zodiac ring is painted blue, with symbols and names of the individual signs in ochre. The sun, represented by a boxwood sphere, is carried by a brass rod passing through the poles, the earth and moon, being carried on a subsidiary arm from the same rod. The sphere is held in a red-painted meridian ring slotting into a horizon ring of iron covered with paper bearing a degree scale and zodiacal calendar. The whole is mounted on a three-footed stand in *neo-greque* style, with applied gilt-brass *oiseau* decoration and claw feet, on a triangular red varnished wood base, with inset silvered compass.

Fig 35 *(overleaf)*
Indian astrolabe, eighteenth century.

Fig 36 *(overleaf)*
Islamic astrolabe, signed Muhammad b. Hamudî al-Isfahânî, 1175/6.

Fig 37 *(overleaf)*
Persian astrolabe, signed Muhsin b. 'alî as-sharif, 1780.

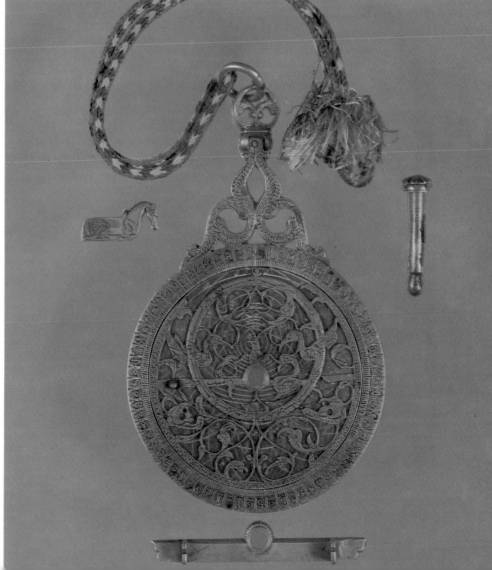

Fig 38
German planetary volvelles, *c.*
1574–5, printed paper with three
rotatable 'dials', mounted in pine
frames.

Volvelles for the sun and Mercury,
probably taken from the set of eight
published by Leonhardt Thurneisser
*Des Menschen (der sonnen, des
mons . . .) circkel und lauff* for the
years 1575–80.

The two instruments shown here were
designed to ease such astronomical
calculations as were necessary in the
practices of astrology, alchemy and
Paracelsan medicine. The inscription
in the cartouche (on that for the sun),
discusses the exact position of the sun
on the 6th day preceding the Ides of
March (i.e. 10 March), according to
five different chronological systems.
The outer rim of the volvelle itself is
marked '*primum mobile habitatio
Dei altissimmi potentissimmi cum
sanctis electis suis qui est omnium
rerum*'. (The primum mobile, the
habitation of the all-powerful God
most high who is lord of all, with his
elect spirits.) The second ring contains
the names of the more important
angels, and the third (wider) ring
contains a calendar scale with various
named festivals. Inside this is a
divided ring containing the names of
various countries with their
latitudes, and following this a ring
containing thirty-two wind names.
The last ring of the base contains a
zodiac scale with twenty-eight named
stars (of Paracelsan significance?).
Engraved in the central circle (only
visible lower) are an *aspectarium*
and a diagram of the celestial houses.

Rotating over and within these scales
are three 'dials'. The lowest of these
shows pictorial representations of
various star constellations; the
second with what appears to be a
circular tree diagram, partly marked
in Latin and partly in high German;
the third carries further star
constellations. Above these rotates an
ecliptic circle, a dragon representing
the nodes, while the centre of this is a
figure of Hermes shown with
attributes which are specifically
Paracelsan. The Aramaic inscription
surrounding this figure consists of
the three words for wisdom,
knowledge and Jehova. The figures
of the lion and the king in the upper
left and right cartouches both
represent gold in alchemical
symbolism, and here probably also
imply the male principle and the
macrocosm. The figure at the lower
right represents the queen and the
two figures on the bottom, of whom
the left is shown blind, represent the
female principle and the microcosm.

The figure of the Paracelsan Mercury.

Constellations of the northern
hemisphere with pictorial zodiac signs.

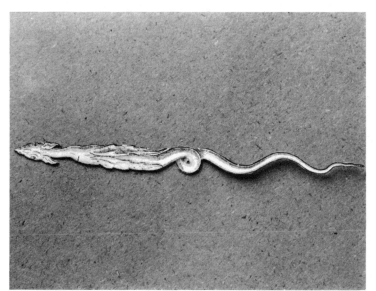

Dragon pointer for the nodes.

Constellations of the southern
hemisphere with plain zodiac.

Sun and Mercury ring.

Fig 39
Individual parts from the planetary
volvelles by Leonhardt Thurneisser,
1574–5.

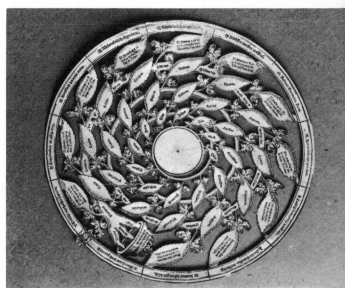

The universal tree.

Fig 40 *(overleaf)*
Celestial and terrestrial globes, by
W. J. Blaeu, seventeenth century.
See also fig. 24.

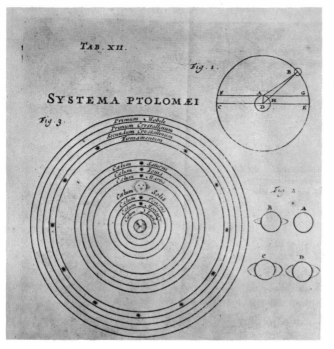

TAB. XII.

Fig. 1.

SYSTEMA PTOLOMÆI

Fig. 3

Fig. 2

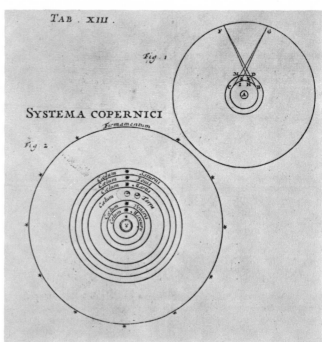

TAB. XIII.

Fig. 1.

SYSTEMA COPERNICI

Fig. 2

Firmamentum

The Sun
The Earth.
The Moon.
Venus
Mercury.

The Orrery

The Ecliptic

Made by M^r. *Iohn Rowley*   *M*^r. *of Mechanicks to His Maj*^{ty}

Orreries

Although the armillary sphere provided a convenient and attractive representation of the Ptolemaic, geocentric, system of the universe, it was not so well suited to show the Copernican, heliocentric, universe. The con-planar nature of the Copernican system, i.e. with all the planets going round the sun in approximately the same plane, meant that a different form of instrument was required, and as a result geared models were devised to show the planets revolving with correct relative velocities one to another.

The first orrery was made by two clockmakers, Thomas Tompion and George Graham, his nephew by marriage, who produced a clockwork model which reproduced the daily motion of the earth and the periods of the moon. This model, correctly termed a 'tellurium' made of silver and ebony, was traditionally supposed to have been made for Queen Anne to present to Prince Eugene of Savoy. In 1710–12 it was seen by John Rowley who copied the instrument for his patron, Charles Boyle, the fourth Earl of Cork and Orrery. According to Richard Steele, Rowley called the machine after his patron, and, although James Ferguson in 1746 protested, the name 'orrery' stuck to the planetarium. Thomas Wright, who had shared premises with George Graham, developed the tellurium into a complex planetary system of long moving arms bearing the planets and their satellites, which was termed the grand orrery.

Orreries were expensive furnishings, and the need for a more modest instrument which would sell at a popular price suitable for schoolrooms was met by a range from Benjamin Martin after 1766 (fig. 47). Also made by both George Adams Jnr and Dudley Adams (fig. 52) they achieved an elegance and beauty that was unsurpassed. W. & S. Jones, T. Blunt (fig. 48) and Newton also made instruments in this form, although it is still not clear whether each of them made all of the parts, or whether they were subcontracted. W. & S. Jones were manufacturers of considerable size and it is quite possible that they made a great number of parts, if not all, of many of those bearing other names. Such was the demand for orreries that by about 1814 W. & S. Jones could list in their catalogue a whole range of orreries, in all varieties from the simple to the *de luxe*, from £40 to £1,000. These brass and steel models were originally made to fold away and be stored in a mahogany case. The circular silvered calendar plate with a worm-wheeled edge to take the alternative lunarium and tellurium attachments is engraved with a compass-rose at the centre and a zodiacal scale on the circumference.

The basic form of an orrery was a representation of the sun and planets on the same plane showing their relative motions rather than their sizes and distances from each other. Arms of different lengths radiate out from a central shaft, bearing a gilt brass sphere representing the sun, and carry ivory planets, with their attendant moons. When the mechanism is set in motion, which is done with an ivory-handled crank, the planets move in ratio to each other. It would have been impossible to reproduce the relative sizes of the planets in a portable instrument, for if the earth were scaled down to the diameter of a 3 inch pocket globe, the accurate relative size of the sun would be about 26 feet.

With the increasing demand for even cheaper instruments, W. & S. Jones made a series of orreries which were stripped of all expensive refinements, such as hand engraving on brass and silver, and used simple materials like wood and paper. They have hand-coloured printed calendar plates which are full of information and are mounted onto a wooden base. The size and number of radial arms bearing ivory planets depended upon the model, but on each they were powered by an endless screw which turned exposed gear wheels underneath the base (figs. 50, 51, 53).

William Jones described his portable orrery in 1812 in his *Description and Use of a New Portable Orrery*:

'The board on which the machines are turned, has pasted on it a printed paper, coloured and varnished; near the outward edge of which are several circles: the first or outward circle is divided into the twelve

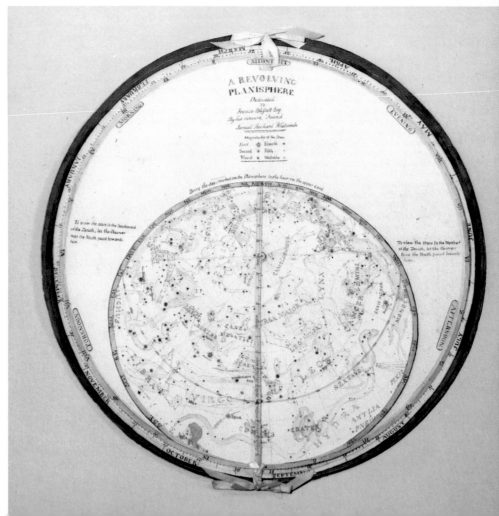

Fig 47 *above*
English orrery signed '*B. Martin, Inv'. et Fecit Londini*'; between 1765 and 1782; brass with ivory spheres for planets, globe for the tellurium, attachment of printed gores on a core, coloured and varnished.

The cylindrical drum containing the mechanism is mounted on a single column stand with three folding cabriole-type legs. Attached to the top of the drum is a silvered ring engraved with a zodiac/calendar scale toothed on its outer edge. In the centre is a compass-rose with unnamed points. Central shaft for sun to which the planetarium, lunarium or tellurium fittings may in turn be attached. Ebony handled crank. The planetarium fitting has planets to Jupiter. Earth having one satellite moon, Saturn five and Jupiter four. Benjamin Martin (1784/5–1782) was largely responsible for the development and popularization of the double-core planetarium in an attempt to produce a cheaper form of the large and extremely expensive instruments of the type made by John Rowley and his successor Thomas Wright (see fig. 42, and John R. Milburn, 'Benjamin Martin and the Development of the Orrery', *British Journal For the History of Science*, vi, 24, December 1973, pp. 378–99).

Fig 48
English orrery, signed 'T: Blunt London', globe signed 'Lane's Pocket Globe London, 1809', *c.* 1810. Diameter of calendar plate 225 mm ($8\frac{7}{8}$ in.), overall height 473 mm ($18\frac{5}{8}$ in.). Brass and steel; globe of printed and coloured paper gores on a (?) plaster core.

The body of the instrument consists of a silvered and toothed calendar-plate, with a 12-point compass rose at the centre, and a zodiacal calendar at the circumference. This calendar-plate is the top of a cylindrical box which supplies the base of the instrument, being supported on a single brass column carried by three folding cabriole-type legs. Inside the base is a geared mechanism, operated by a detachable, ivory-handled, crank. This, when turned, allows the concentric planetary arms which may be attached to the central shaft, to be rotated. A brass sphere to represent the sun may be attached to this central shaft. Alternatively the lunarium or tellurium fitments may be attached.

The planetary arms, with ivory spheres for the planets, are in order outwards from the sun: Mercury, Venus, Earth with moon, Mars, Jupiter with four satellites, Saturn with its ring and six satellites, Uranus.

At a later date the central shaft has been modified to carry two further planet arms.

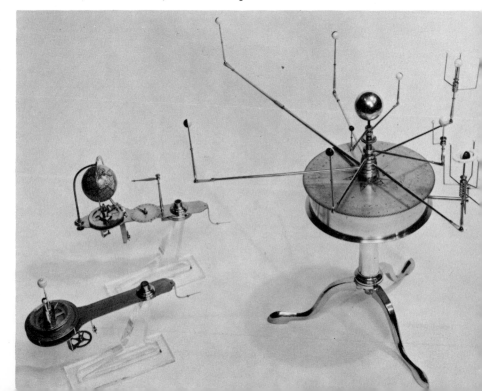

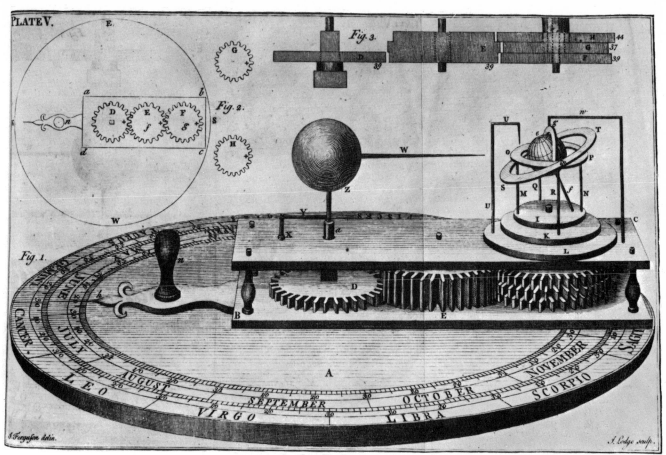

Fig 49 *above*
James Ferguson's, 'Mechanical
Paradox', engraving on paper signed
'J. Ferguson delin.' and 'J. Lodge
Sculp.', from *Select Mechanical
Exercises: showing how to construct
different clocks, orreries, and sun-
dials, on plain and easy principles* ...
London 1773, plate V, between pages
44 and 45.

The 'paradox' of Ferguson's machine
was that the teeth of one wheel
meshing to equal depths in three
others should so affect them 'that
in turning it any way round its axis it
should turn one of them the same way,
another the contrary way, and the

third no way at all'. In the
illustration it is shown arranged as a
tellurium, i.e. displaying the
phenomena of seasons, day and
night, lunar motions and times when
eclipses will occur.

Fig 50
English orrery, signed 'Designed for
the New PORTABLE ORRERIES by
W. Jones and made and sold by
W. & S. JONES 30 Holborn, LONDON', *c.*
1810—20, diameter approx 315 mm,
($12\frac{1}{2}$ in.).

Circular base covered with a paper
printed round the edge with a zodiac
calendar. Contained within this in the
upper half of the circle is 'A TABLE of
the principal AFFECTIONS of the PLANETS
published by W. & S. Jones in 1794',
showing the distances, periods, sizes,
etc. of the planets out to Saturn. In the
lower half of the circle is a pictorial
representation of the solar system. A
brass sphere representing the sun is
mounted on the central shaft around
which revolve the inferior planets
represented by ivory spheres, and the
earth and moon. The moon is mounted
over a silvered zodiac scale, with a
silvered dial carrying a lunar phase
diagram below. The whole machine
is operated by turning a cranked
handle which may be attached to an
arbor below the base-board, meshing
by an endless screw to a wheel
attached to the central shaft of the
movement which is carried through
the base.

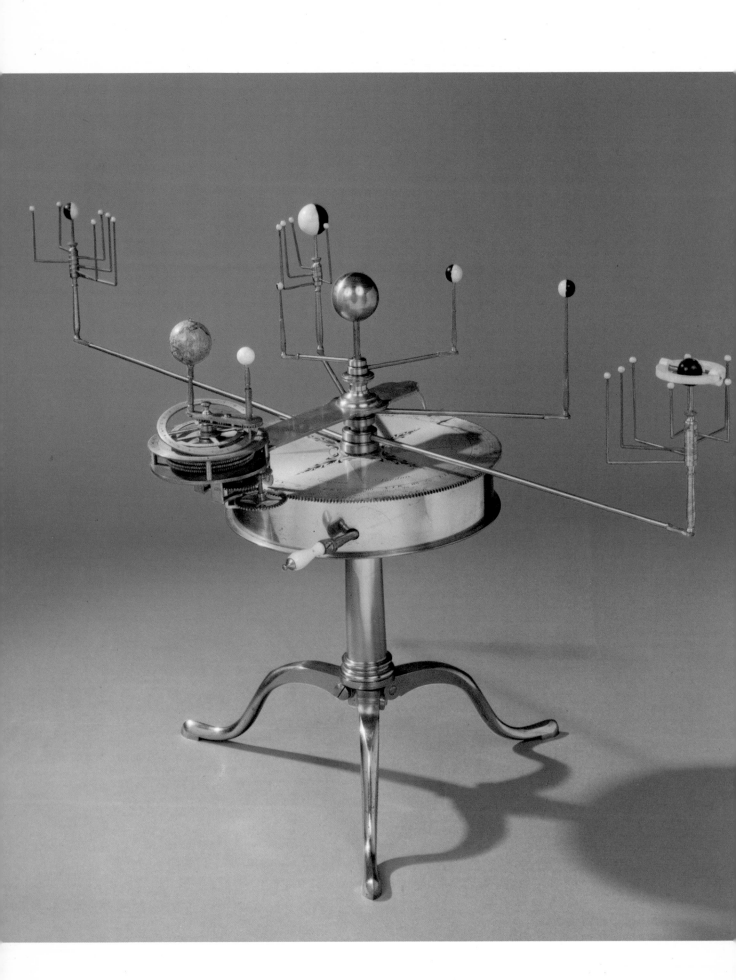

Fig 51 *right*
English orrery, signed 'A NEW PORTABLE
ORRERY Invented and Made by
W. JONES and sold by him in Holborn,
LONDON' *c.* 1812

Movement of brass and steel;
mahogany base covered with printed
paper, coloured and varnished;
planetary arms of brass with ivory
spheres for the planets; two carrying-
cases of mahogany.

Circular base overlaid with paper
printed with a zodiac calendar and the
four seasons together with the length
of day and night and the equinoxes.
In the central circle are pictures of the
planets each drawn in a 'proportion as
near to each other as possible, and
that of a globe of the diameter of the
board for the sun'. The instrument
may be assembled either as a
tellurium (i.e. to show the
relationship of earth and moon
around the sun) or as a planetarium.
For use as the latter, six arms which
are attached by collars to the central
axis carrying the sun may be put in
place. The planets shown are Mercury,
Venus, earth with moon, Mars,
Jupiter with four satellites, Saturn
with ring and five satellites. A small
brass ball which was used to
represent the sun if the instrument
was set up to show the Ptolemaic
system is now missing.

For use as a tellurium, a frame
containing four wheels, earth, moon
and sun is screwed into place on the
central axis. The earth with its axis
inclined at $23\frac{1}{2}°$ is engraved with the
arctic, antarctic and tropical circles
the equator and twenty-four
meridians. It is surmounted by a
'terminator', as Jones called it, to
show the boundaries of light and dark.
The moon is geared to move round the
earth in its period of $29\frac{1}{2}$ days, and is
given its proper inclination by the
hinged ring below. Beneath this is an
ecliptic ring for showing the moon's
position, and below this a card
showing the lunar age and phases.

Engraved on the tellurium frame is
the inscription 'Monthly Preceptor
No 2 To Miss Eliz[th] Parker Aged 14 of
Mettingham, near Bungay Suffolk as
the Reward of distinguished Merit'.

The planetarium attachment is
contained in its own mahogany box
which fits with the instrument into
the outer case. Accompanying the
instrument is a copy of the original
instruction book, *The Description and
use of a New Portable Orrery on a
simple construction representing the
motions and phenomena of the
planetary system to which is
prefixed a short account of the solar
system* sixth edition, with additions,
by William Jones, London 1812.

Fig 52 *left*
Orrery by Adams, late
eighteenth century.

months of the year, and the rest into their respective days; the two
innermost circles have the twelve signs of the Zodiac, with their proper
names and characters each sign contained 30 degrees, or the whole
360°: this ecliptic circle represents that path in the Heavens which the
Sun seems to describe in a year, though being properly the path of the
earth round the sun. . . . In this Ecliptic circle are four letters N.W.S.E.
signifying the North, West, South and East parts of the Heavens, as
seen from the earth.

It has also four divisions made by lines drawn from the first degree of
Cancer, the first degree of Capricorn, the first degree of Aries and the
first degree of Libra; between these four lines are inserted the seasons of
the year, the lines denoting their beginning, duration and end; and
along these four lines are inserted the length of day and night, and the
distinction of the Equinoxes.'

Several astronomical phenomena may be demonstrated with this instrument. When the large brass ball is placed in the centre, and the planets in
order of orbit around it, the Copernican system is reconstructed. By
placing an alternative small brass ball in place of the ivory one which
represents the planet earth, and placing earth in the centre, the Ptolomaic
system is portrayed. When a small oil lamp is provided with the outfit
this can be placed in the centre in the sun's place and when it is lit, and the
planetarium is set in front of a white wall, 'by moving the satellites of
Jupiter and Saturn round the planets, the projection of their shadows will
shew the reason why those moons always appear on each side of Jupiter
and Saturn in a right line; why those which are most remote appear
oftentimes the nearest; and *vice versa*'. Mr Jones modestly declared that
his orrery made possible the general comprehension of astronomy, which
is more than likely, for the problems of three-dimensional quantities are
extremely difficult to visualize without a diagram.

The orrery could be arranged according to the current positions of the
planets from the *Ephemeris*, the diary or daily account of the movements of
the planets, which, now published by H.M. Stationery Office in the
Nautical Almanac, was originally published privately. The *Ephemeris* gives
the positions as they would be seen throughout the year from the sun or
Earth, termed heliocentric or geocentric places. In such a publication the
information on how to find the place is listed as angular distance from the
other planets, and the constellations of the Zodiac (fig. 57).

Fig 53
English orrery signed 'Designed for the NEW PORTABLE ORRERY by W. JONES and Made and Sold by W. & S. Jones, 30 Holborn, London', 1810–20, diameter approx. 315 mm (12½ in.).

The more elegant table model of Jones's New Portable Orrery. Uranus (not shown) has six satellites following Sir William Herschel's report in 1798 of four satellites additional to the two he had previously discovered. This form of the New Portable was listed in Jones's 1812 catalogue at prices between £16 16s and £22 1s.

Fig 54 *left*
French planetarium, signed on the globe ' G[LOBE]T[ERRESTRE] FOR [TIN] ... PA[RIS]', c. 1800, diameter 500 mm (19¾ in.), wood and brass; printed coloured and varnished. Stand painted black with gilt floral decoration on the edges; the rims painted red.

Turned base which contains the mechanism actuated by rotating a turned ivory handle (a later replacement). The planetary arms are attached to a central turned wooden column by brass collars. Gilt globe representing the sun, around which rotates the earth, the moon and the planets out to Saturn. Each planet, except the earth, which is represented by a globe, is shown by a printed paper emblem. The whole is surrounded by an octagonal ring overlaid with a printed paper carrying a zodiacal calendar with pictorial representations of the signs of the zodiac.

Fig 55 *right*
English engraving of an armillary orrery inscribed 'A Perspective View of Mr H[a]wk[e]s's Orrery'. Printed for J. Hinton in Newgate Street and published in the *Universal Magazine*, September 1755.

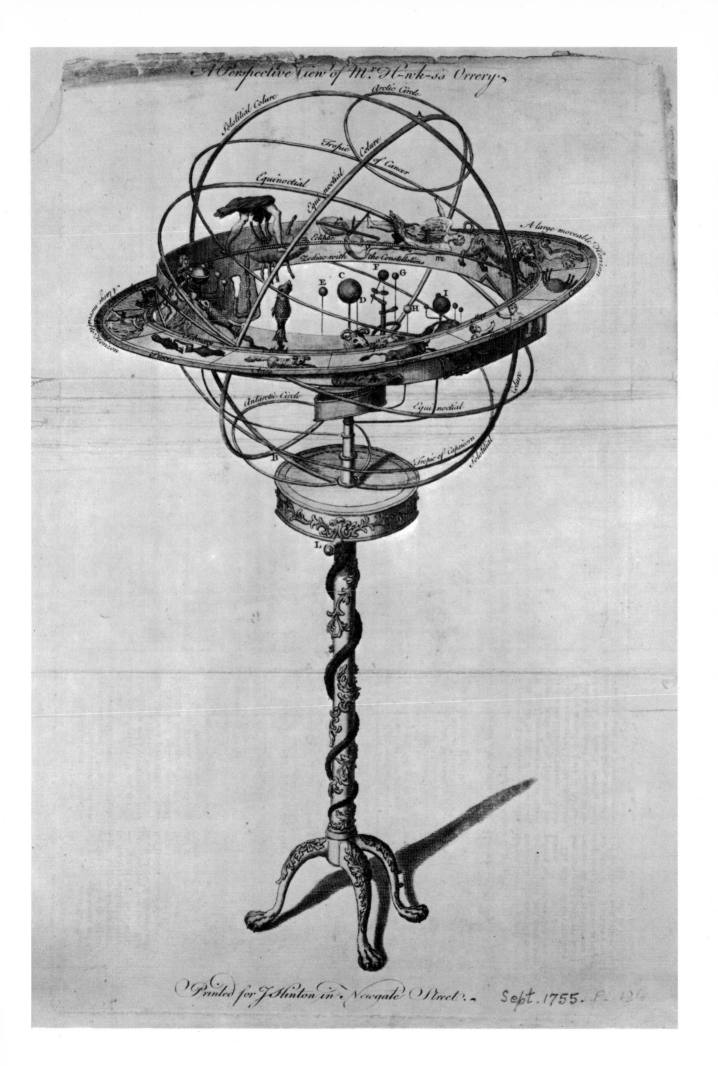

A Perspective View of M.^r H—nk—s's Orrery.

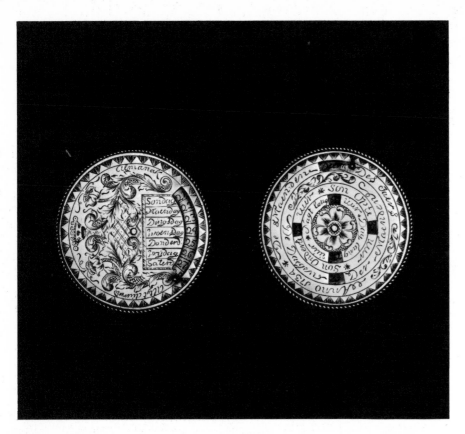

Navigation ²

IN the words of Dr John Dee, Elizabethan scholar and tutor of navigation, written in 1570 'the art of Navigation demonstrateth how by the shortest good way, by the aptest direction and in the shortest time a sufficient ship . . . be conducted' (quoted by Waters *Navigation*, p.3). The problems were manifold, and it is how these problems were resolved that is so fascinating.

Two main tasks confronted the navigator, to plot a course and to determine his position. Coastal navigation depended on pilots and their accumulated knowledge of local conditions, tides and high water, the use of the compass, and the lead and line, but oceanic navigation depended on the determination of two co-ordinates, latitude and longitude, the so-called 'Masterpiece of Nautical Science'.

The *primum mobile* was trade. Self interest caused the accumulation of knowledge to be jealously guarded at first, and could explain why Raleigh employed Thomas Hariot as his personal mathematician, and why Drake carried his personal chartmaker aboard, who sketched views and coastlines for future expeditions. It could also explain why he made a particular point of capturing the pilots of foreign ships, seizing their charts, sailing directions and instruments, to add to his store of information of coasts on both sides of the Atlantic for his circumnavigation of the globe in 1577–80.

Gradually, the dissemination of scientific discoveries enabled those who were capable, if not to master the sea, at least to contribute actively towards finding their destination.

Fig 58
English(?) lodestone, not signed,
c. 1630, 60 × 34 × 27 mm
(2½ × 1⅜ × 1⅛ in.).

Brass-bound lodestone with simple
brass suspension ring. Decorated with
double scribed vertical and
horizontal lines.

Fig 60
Italian compass, not signed, c. 1650,
diameter 70 mm (2¾ in.), wood.
Turned and glazed box with
pivoted 32-point compass card
painted by hand. The north point is
indicated by a star, the east by a cross.

Direction-finding

Fig 59
English trade label, signed 'Thomas
Lorkin Mathematical Instrument
Maker, N° 89, near New Crane Stairs,
(WAPPING) London', c. 1800.

An attractive example of a maker's
label, which supplies the only known
evidence (at present) of his
existence. Labels of this kind do not
supply good evidence for the date of
instruments in which the makers
who used them dealt from the
illustrations, but they are a useful
source for addresses and sometimes
the stock when, as here, a written list
of items supplied is given.

The Compass
The magnetic compass as we know it depends upon two things, the lodestone, a naturally occurring mineral with magnetic properties, and the ancient custom of direction-finding by the North star.

The Greeks were familiar with the lodestone, a form of soft iron capable of creating a magnetic field around another piece of iron, 600 years before Christ, and the Romans knew about the properties of magnets. Claims were once made that Amalfi in Italy invented the compass, mainly because of its importance as a sea power between the sixth and twelfth centuries, when its land routes had been blocked, and because William of Puglia wrote in 1109–11 that she was 'famous for showing sailors the parts of the sea and the sky' *(Gestes de Guiscard 'Nauta Maris Coelique Vias Aperire Peritus')*. These, however, have been discounted by later writers.

The Chinese are known to have had a south-pointing carriage which had nothing whatsoever to do with magnets, but they did have what they termed a 'point south' – south being a venerable direction in Chinese philosophy, which was a magnetized needle, and there are records that it was in use at sea in A.D. 1101 and 1126. The Arabs transmitted what tenuous contact the Chinese had with the West, but there is no record from them of the Chinese device. The earliest known record of an Arab compass was in A.D. 1243 and as the Arabs had inherited a great deal of their knowledge including magnetism from the Greeks, it was probably a case of two inventions being made quite independently.

The magnetized needle must have been known to northern mariners by the twelfth century for the first known mention of it was made in 1187 by Alexander Neckham, an English monk who studied in Paris. He described a needle transfixing a piece of reed and floating in a bowl of water, and he also described a pivoted needle, 'a needle placed upon a dart', which whirled around until it pointed in a northerly direction. By 1218 a compass used to check the direction of the wind in bad weather was deemed 'most necessary for such as sail the sea'. The magnetized needle floating in a bowl of water does not seem to be a very satisfactory scientific instrument in a small ship as any amateur yachtsman will agree, so it is most unlikely that its use at sea was widespread.

The first description of a dry compass, a magnetized needle mounted on a vertical axis in a wooden bowl, is found in a treatise on magnets by Pierre de Maricourt (Petrus Peregrinus) a French soldier, in 1269. He described how the bowl was fitted with a graduated verge ring and a non-magnetic pin was fitted at right angles to the needle so that the moments of inertia in the two vertical planes passing through the needle at right angles to each other were equalized. Without it the compass would be unusable in a seaway on board a pitching and rolling vessel, for the needle would tend to turn into the plane of the ship's roll.

The familiar compass card, termed a compass-rose or compass fly, with directions taken from the cardinal points, became attached to the magnetized needle, which rotated independently for direction finding, and was enclosed in a box at some time about the end of the thirteenth or beginning of the fourteenth century.

In ancient times the method of direction-finding was with the winds, roughly aligned from the position of the Pole star, which had long been known to be the only star in the firmament that did not move. Its reliability as an orientation aid was recited in poem and saga. It was an easy step on land to find south from the mid-day position of the sun, and the shadows it cast served to indicate east and west. These were not exact cardinal points as we understand them, but a general arc of direction, for in the Mediterranean, should a sailor make a landfall off course, he had only to stop and ask the way. From Homer's time these four directions were named after winds which blew the square sail vessels on a predetermined course, and were referred to as 'following winds'. It was a logical step to increase the intermediate directions with further wind names. It is important to note that the wind directions on early charts did not refer to universal cardinal points, however vague, but to the relative directions according to which area the chart represented. These wind directions, depicted in the form of a spoked wheel, which appear on early charts, are termed wind-roses and are not related to compass-roses.

The Italians identified the north wind with the Tramontana mountains, which were due north according to the Pole star; the other wind names betray their origin: *Grecho* (N.E.), *Levante* (E.), *Sirocco* (S.E.), *Mezzodi* or *Ostra* (S.), *Garbino*, *Libeccio* or *Africus* (S.W.), *Ponente* (W.) and *Maestro* (N.W.). These names or their initials are frequently used on Italian and sometimes on South German (Augsburg) instruments instead of the more familiar cardinal points, which were adopted from the Frankish initials. On the compass-rose, the north was represented by the *fleur de lys* from about 1500 and until the eighteenth century the east by a cross.

By the fifteenth century confidence in the accuracy of the compass to take magnetic bearings of ships and objects was such that ship masters changed course by its bearing, in preference to the older method of by the trim of the sails.

Originally painted in colours the compass fly was later painted black on a white card or white on a black card, so that the direction lines, or 'rhumbs', would be clearer at night in the light of the candle lantern used for illumination. Great care was taken to line up the 'lubber' (or lubber's) line marked on the verge ring or in the bowl of the compass, with the fore and aft line of the vessel.

The compass box was made of wood turned out of the solid, half the diameter of the fly in height, covered with glass, sealed in with resin. The base could be removed and on its upper side the pivot or brass pin could be fixed to take the fly. The compass 'wire', or needle, was a length of soft iron wire, equal to the circumference of the fly, bent double into a loop, and pinched at each end to equal the diameter of the fly, then stuck onto the back of the fly with paper. A brass cone, or 'capital', was pushed through the centre of the fly, through the wire loop, onto which the fly could pivot. On a long voyage the soft iron wires would need frequent 'refreshing' (magnetization with the lodestone) which, 'armed' with a soft iron keep-plate to produce the best 'virtue', was removed from its case and the iron wires stroked with it 'as you would whet a knife'. The steering, or

Fig 61
English ship's compass, c. 1850, not signed, diameter 159 mm (6¼ in.), brass.

32-point compass gimballed in a small wooden binacle.

'conning', compass was mounted in gimbals into another box. Gimbals, from the Old French *gemel* (twin), were two brass rings, which moved within each other in such a way that despite the movement of the vessel, the compass was held level between them. Gimbals have been in use since the sixteenth century and were mentioned by Cortes in 1545 and Pedro Nuñez in 1537.

The phenomenon of 'variation' became apparent from floods of observations taken from 1492–1500 when long voyages to America and India began. Seamen were terrified by the apparently supernatural behaviour of the needle, which in spite of their earlier distrust of what had appeared to be magic, they had now been trained to obey implicitly. From the time of Columbus onwards compass variation according to the distance travelled in a westerly direction was observed, due, as we now know, to the magnetic forces of the earth. It was recorded on charts, and it was used as a kind of guide to longitude, although some writers suggested that the lodestone was at fault. In the meantime, instrument-makers in Flanders made a correction to the compass by re-aligning the fly over the needle so that when the compass indicated magnetic north, it was really showing true north. This caused a great deal of confusion and added in no small part to the hazards of the sea. A compass of variation evolved when an alidade was attached over the box so that a bearing could be taken to check the azimuth of the compass with the Pole star. 'Deviation', error induced in the compass by the presence of iron, while not described, was understood in the sixteenth and seventeenth centuries, as is attested by the use of only wooden pegs in the construction of the binnacle. When variation was first observed, seamen checked to see if they had knives or other metal objects about the compass to account for its behaviour. During the seventeenth and eighteenth centuries the demand for instruments increased and caused a number of slip-shod compasses to be made with iron nails and careless balance, about which there were many bitter complaints.

Navigators devised methods of checking the compass. 'The pilot's blessing' was familiar to sixteenth-century sailors who would see him in the twilight high up on the half deck aft with his hand stretched out to heaven as if he were blessing the ship. What he was really doing was pointing to the Pole star and letting the plumb line drop to the compass, so as to observe any error – what we should today term a 'hand bearing'. He would also be checking the correction to altitude necessary for latitude determination because the Pole star is not exactly at the magnetic pole, as will be explained later. In order to do this the lead-line was extended to cut the Pole star, Casseopeia and the Great Bear, to check whether the Pole star was exactly below or above the pole.

While variation occurs in different longitudes, deviation depends upon the direction of the ship's head. Sailors are familiar with 'swinging the ship', turning the ship's head in several directions so that a deviation scale can be devised. Compass adjusters compare the compass bearing of the sun with its known correct bearing at the time, or use a pelorus (fig. 201) which is a compass fitted with a lubber line and rotatable sighting arms, with which to check the ship's compasses.

With the advent of iron ships in the nineteenth century the binnacle was developed into an all-brass fixture, which surrounded the compass with a protective non-ferrous overcoat. These splendid survivors of Victorian England are rather large for an intimate collection, but a small 'lifeboat binnacle' is in fact still being made, and is statutory equipment on ships' lifeboats.

Seventeenth- and eighteenth-century dry card compasses are occasionally to be found (fig. 60) but tell-tale compasses are rarer. These were placed overhead in the cabin and for this reason the cardinal points are reversed (fig. 62). It must have been a happy discovery that alcohol did not freeze as readily as water, although Leonard Cushee noted on his terrestrial globe *c*. 1750 that even brandy froze towards the north pole. The compass bowl made of brass, originally filled with spirits of wine, was

Fig 62
English 'tell-tale' compass, not signed,
late nineteenth century. Diameter of
bezel 130 mm (5⅛ in.). Printed 8-point
compass card with *fleur de lys*
indicating the north point. In circular
silver-plated box in a gimbal mount.

Fig 63
Scottish ship's binnacle, signed
'LORD KELVIN'S PATENTS (SIR WILLIAM
THOMPSON) No 7068 JAMES WHITE SOLE
MAKER 46 CAMBRIDGE ST GLASGOW', *c.* 1895,
wood, brass and steel with printed
paper compass card.

Hollow cylinder on square base which
may be attached to the floor, and
incorporating a plumb level signed
'J. WHITE GLASGOW'. Inside the
cylinder are racks for the corrector
magnets. Brass cap with observation
window for the gimballed, heavy card
compass. On the exterior of the case
are two brackets carrying soft iron
spheres.

Lord Kelvin's patent compass and
binnacle, while not employing any
new design principles, revolutionized
compass manufacture, being the first
to combine successfully all the
requirements for an accurate
instrument in one model. From the
time of its introduction in the late
1870s, Kelvin's binnacle provided the
basis of all subsequent British
developments.

an eighteenth-century development when artificial magnets which did not require refreshing proved to be reliable, although the dry card compass was resuscitated by Sir William Thomson, later Lord Kelvin, in 1876, and was in general use in the merchant navies of the world well into the twentieth century.

Charts and chartwork

Until the sixteenth century charts for local areas, which had developed out of the Roman *periplus* and Italian *portolan*, were drawn on the assumption that the world was flat. These were termed 'plane sailing' charts, incorrectly spelled by the English 'plain sailing', giving us the colloquial expression for something which is simple and straightforward. As we have seen, mariners were simple men, and inclined to follow the sea in the traditional manner of their fathers. Until navigation became a complex art sailors relied on the rutter and the almanac. The English rutter, from the French *routier* which had been in use since about the fourteenth century, was a much-used handbook for limited areas, containing information on tides, entries into ports and harbours, signs and tokens of the sun, moon and stars at various times and seasons, and the look and sound of the

Fig 64
Title page of the third edition (*c.* 1590) of Robert Norman's influential Rutter, the first in English to be illustrated with wood-cuts of the coasts. Norman's text was based on two Dutch works. The main part derives from Cornelis Anthoniszoon's *Leeskaart boeck van Wisbury* 1566.

Fig 65
Title page of the first British translation of W. J. Blaeu's *Eerst deel der Seespiegel, inhoudende een karte onderwysinghe inde* Amsterdam 1623–7. In its three main parts the work provided a primer of navigation, a waggoner (i.e. a set of sailing directions with charts and elevations of coasts) for the North Sea, Baltic Sea and Arctic Ocean, and a waggoner for Atlantic voyages to the Straits of Gibraltar, the Canaries and the Azores. In all *The Sea-Mirror* provided a summary of the most recent navigational information for north-west Europe. It also illustrated the dependence upon Dutch work on such matters prevailing in north-west Europe at the time.

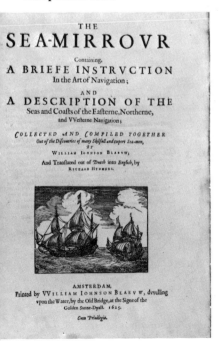

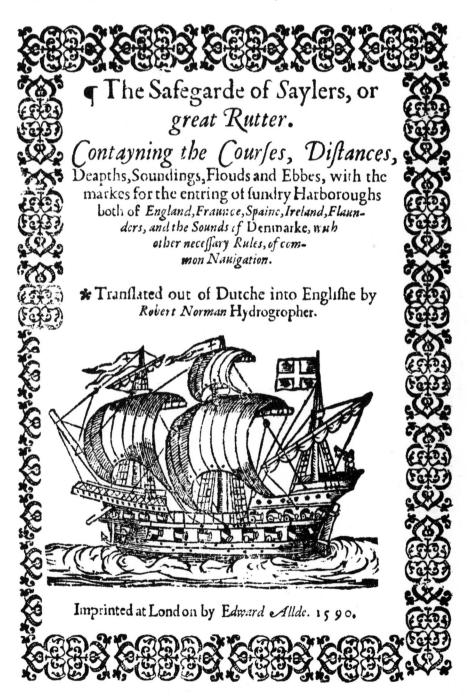

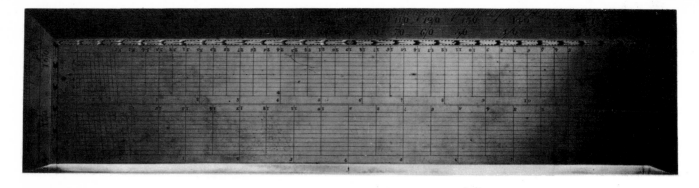

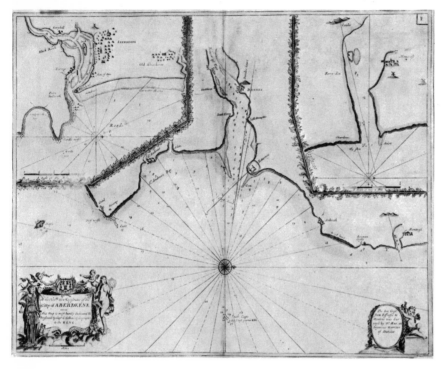

Fig 66
English protractor and plotting or
diagonal scale, not signed, *c.* 1740,
303 × 72 mm (12 × 2¾ in.), brass.

A double diagonal scale used for
dividing a given nominal quantity into
100 equal parts.

Fig 67
English chart from Greenvile Collins
*Great Britain's coasting pilot, being a
new survey of the sea coast* 1693,
555 × 440 mm (21⅝ × 17¼ in.). Hand-
coloured chart, printed on paper.
Collins's survey of the coasts of Great
Britain began in 1681 and lasted for
seven years. The present chart is the
only one for which he acknowledged
any local assistance. Far surpassing
previous work, Collins's atlas was
re-issued fifteen times in the
eighteen century.

sea. Almanacs were in wide circulation by the end of the fifteenth century
and contained a calendar and description of the universe after the Ptolomaic system. There were two types of almanac, one for astronomers,
physicians and scholars, and another type for humbler folk like seamen.
The latter was frequently printed in broadsheet form, and tacked to the
bulkhead, or in pocket-book form for richer detail about moonlit or dark
nights, weather forecast, and the chief astronomical events on which
terrestrial events depended – for this was the age when it was logical to
accept that if the earth were the centre of the universe, as was seriously
believed, then all that was happening in the universe about the earth
would directly affect the destiny of all that was in, of or on it. By the sixteenth and seventeenth centuries such almanacs were owned by almost
everyone, and were rich in all kinds of information, not only astrology
forecasts, but also ecclesiastical data, road distances, details of country
fairs, suitable days for purging, phlebotomy and bathing. Above all they
were for the mariner, containing tide tables, rules for finding high water,
for finding the time by the stars, and especially the phases of the moon,
which were the only source of this information for the sailor until special
tables were printed for him.

Steeped in this mixture of wisdom, superstition and folk-lore, the evolution of the geometrical seaman seemed well nigh impossible. Astronomers
and other scholars who wrote for the seaman presupposed ideal conditions
and there were very few ship masters who could adapt them to real conditions aboard ship.

It was generally known that the earth was not flat, but the plane charts
were based on this false premise, and were faulty, a fact first pointed out by

Pedro Nuñez in 1537. The charts showed the meridians parallel and equidistant from the north to the south pole, instead of converging on these two points, as they do on a globe.

The problem of how to represent the globe on a flat surface was resolved by Gerhard Mercator (1512–94) in his world map of 1569. His solution to the converging meridians was to keep them equidistant, but stretch the latitude scales between the meridians as they recede from the equator. However, it was not easy to use. Loxodromes had always been shown on plane charts as straight lines, but as the earth is a sphere these really should be arcs if they are to represent the shortest route between two places. Mercator devised a mathematical calculation for the mariner involving the multiplication of their departure (the distance run in an east-west direction) by the secant of their latitude ($\chi \times ls$) to arrive at their longitude, but it was far too complicated for him to use.

Although Mercator explained how to use his projection, he did not describe how he had calculated it, and in spite of the mathematical explanations of Edward Wright in 1599 to all who would listen, it was a long time before Mercator sailing was actively employed. Some mariners used both plane and Mercator, although it was considered aboard men-o'-war that Mercator was more accurate. It is not really clear *how* they navigated, for there seemed to be no universal system. Mercator sailing still needed to be championed in 1722, by Thomas Haselden, and there are records that East Indiamen even as late as 1750 clung to the plane sailing method, although the prevailing winds dictated the course the mariner steered to arrive at his destination. The Portuguese, in their attempt to find a fast route to South Africa, found it expedient to sail southwestwards to the Brazilian coast (thus discovering Brazil) in order to pick up the following wind to the Cape of Good Hope.

The tedious mathematical process of transferring a great-circle course from a globe (on which it could not be plotted) to the Mercator planisphere, and the corrections involved, were beyond the capacity of most mariners. The simplification used today is based on gnomonic projection. Transferred to the Mercator chart, the straight rhumb line becomes a curve cutting each meridian at a different angle. The straight line of the gnomonic chart becomes the great circle arc of the Mercator. Gnomonic projection is a projection against a tangent plane, with the point from which the projecting lines are drawn being situated at the centre of the sphere.

The first British Hydrographical Department was set up by the Admiralty in 1795 to assist navigation, sixteen years after the death of James Cook, and the first catalogue of Admiralty Charts was issued in 1825 comprising 736 charts. The United States Hydrographic Office bought the copyright of the almanacs and navigation manuals of Nathaniel Bowdich in 1866 (fig. 69) and have since continued to issue information in connection with the latest developments for the safety of the seaman and his ship.

Although trigonometry had been known since classical Greek times, it was not until 1583 that trigonometrical ratios were set out in tabular form suitable for navigation. Until that time mariners were obliged to use cumbersome long multiplication and division to calculate their own ratios. The invention in 1614 by John Napier (1550–1617) of logarithms simplified the calculations, but as the instructions were written in Latin, they were not immediately available. A set of punched numerals on a series of strips of wood or ivory enclosed in a specially made box was used as an alternative to log tables. These were known as Napier's Bones (fig. 68). Edmund Gunter (1581–1626) simplified Napier's logarithms, and introduced in the *Canon Triangulorem* published in 1620, common logarithm tables which consisted of what we should now call log, sine and tangent to seven decimal places. He was the first to introduce the abbreviation 'log a' for the logarithm of a, and was the first person to use the terms co-sine, co-tangent and co-secant for the sine, tangent and secant of the complement of the arc.

Fig 68
Dutch 'Napier's Bones', signed
'M Dominicus Fecit Amst 1770',
120 × 98 × 22 mm (4¾ × 3⅞ × ⅞ in.),
boxwood and oak.
Ten calculating rods and one double
master.

THE
NEW AMERICAN
PRACTICAL NAVIGATOR:
BEING AN
EPITOME OF NAVIGATION;
CONTAINING ALL THE TABLES NECESSARY TO BE USED WITH THE
NAUTICAL ALMANAC,
IN DETERMINING
THE LATITUDE, AND THE LONGITUDE
BY
LUNAR OBSERVATIONS:
AND
KEEPING A COMPLETE RECKONING AT SEA:
ILLUSTRATED BY
PROPER RULES AND EXAMPLES:
THE WHOLE EXEMPLIFIED IN A JOURNAL,
KEPT FROM BOSTON TO MADEIRA,
IN WHICH ALL THE RULES OF NAVIGATION ARE INTRODUCED.
ALSO,
THE DEMONSTRATION OF THE USUAL RULES OF TRIGONOMETRY:
PROBLEMS
IN MENSURATION, SURVEYING AND GAUGING:
DICTIONARY OF SEA TERMS:
AND THE MANNER OF
PERFORMING THE MOST USEFUL EVOLUTIONS AT SEA.
WITH
AN APPENDIX,
CONTAINING
METHODS OF CALCULATING ECLIPSES OF THE SUN AND MOON, AND OCCULTATIONS OF THE
FIXED STARS: RULES FOR FINDING THE LONGITUDE OF A PLACE BY OBSERVATIONS
OF ECLIPSES OR OCCULTATIONS: AND A NEW METHOD FOR FINDING THE
LATITUDE BY TWO ALTITUDES.

BY NATHANIEL BOWDITCH, LL. D.
Fellow of the Royal Societies of London, Edinburgh and Dublin: of the American Philosophical
Society, held at Philadelphia: of the American Academy of Arts and Sciences; of the Connecticut
Academy of Arts and Sciences; of the Literary and Philosophical Society of New-York, &c.

SEVENTH STEREOTYPE EDITION.
···◦◉◦···
NEW-YORK:
PUBLISHED BY E. & G. W. BLUNT, PROPRIETORS,
No. 154 WATER-ST. CORNER OF MAIDEN-LANE.
George F. Bunce, Printer

1832.

Fig 69 *left and right*
Title page and facing folding chart of
the Atlantic, from Nathaniel
Bowdich *New American Practical
Navigator* ... seventh stereotype
edition, New York 1832.

Fig 70 *right*
English sector signed '*G. Adams
London*' c. 1760, length of arm 229 mm
(9 in.). Brass, with fitted original
mahogany box lined with green felt,
brass catches and hinges. The
instrument is engraved with the sets
of lines of number, chords, sines,
tangents and secants and polygon
which had been standardized on
English sectors during the early
eighteenth century.

Centre right
English single-handed dividers, not
signed, seventeenth century, length of
arm 190 mm (7½ in.), iron.

Far right
English sector signed 'Is Carver Fecit
1704', length of arm 229 mm (9 in.),
brass.

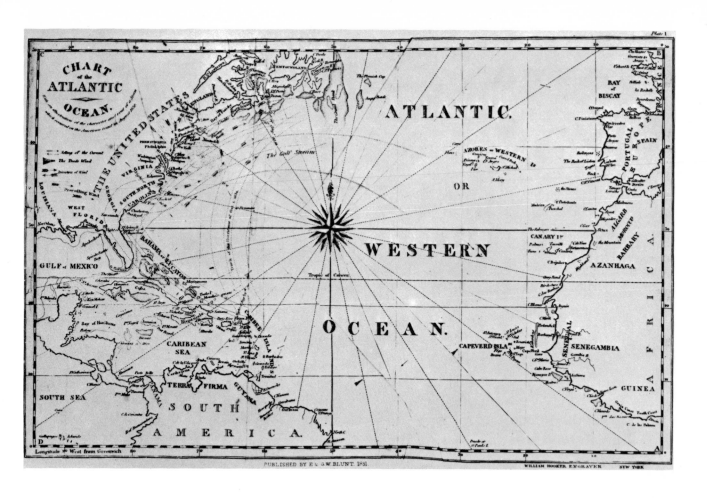

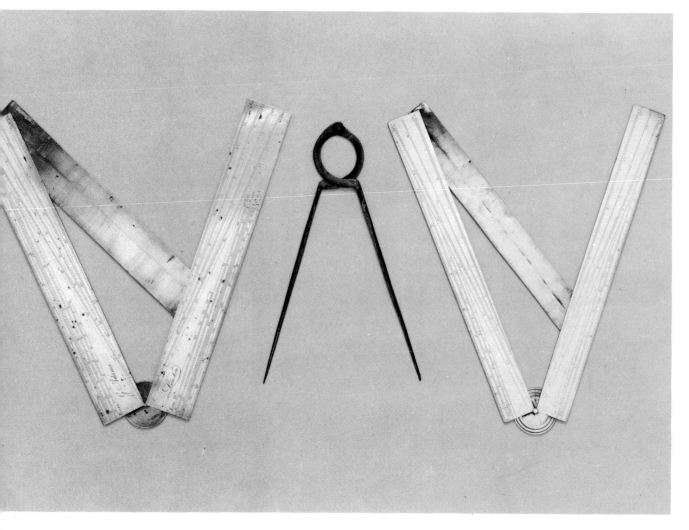

Efforts were made to find a geometrical solution with an instrument which would save the mathematical calculation. Thomas Hood invented a form of sector in the 1590s to be used by surveyors, which was adapted for navigation by Gunter in 1606, and although the instructions for using it were written in Latin and distributed in handwriting, many were made and used by mariners. Gunter published a description in English in 1623 and a year later introduced the Gunter scale, which remained in constant use until the latter part of the nineteenth century.

The sector (figs. 72 and 70) consisted of two limbs made of brass, jointed like a carpenter's rule, on which graduated lines of sines, tangents etc. were engraved. Using this instrument with a pair of dividers, any problem involving right angle triangles or ratios could be solved.

The Gunter scale (figs. 71 and 73) was a straight wooden ruler approximately either 12 inches or 18 inches in length on which log scales and trigonometry functions were engraved. Using this with a pair of dividers the same results could be achieved as with the sector.

Other instruments used for chart-work were the single-handed dividers referred to in old documents as compasses, simple and rolling rulers, and station pointers.

Single-handed dividers (fig. 70) were used for pricking out a course, or measuring distances on the chart against the calibrations of linear measurement on the Mercator latitude scales. The simple ruler became the parallel ruler in 1584 and has changed very little since the beginning of the nineteenth century. A rolling ruler came into use at this time (fig. 74),

Fig 71 *right*
English Gunter rule, signed 'MADE BY WILLIAM PLACE 1694 FOR BEIAMEN PHILLIAPS', 610 mm (24 in.), boxwood.

A good example of a fully signed and dated instrument providing evidence for the existence of an otherwise unknown maker.

Fig 72
English triangular quadrant, not signed but by John Browne, c. 1675–80, 478 × 66 × 11 mm (18¾ × 2⅝ × ½ in.), boxwood and brass. A composite instrument devised by John Browne and described by him in *The description and use of the triangular quadrant . . .* London 1671, which combined scales from the sector, the quadrant and the Gunter rule in one instrument. The example shown here lacks the loose cross piece which fitted by two tenons into mortice-holes at the ends of the two arms of the sector. Also lacking are the plumb line and sights. Browne particularly recommended the 18-inch and longer instruments for use at sea.

Scales on the sector side; the line of lines and line of sines, sharing a common middle line; lines of secants; lines of tangents; sines and versed sines (i.e. co-sines); lines of chords and equal parts.

Scales on the quadrantal side: on the lower arm lower edge a protractor scale of 180°, the divisions being radial from a point at the end of the line of general sines on the upper arm where the angle between the two arms is 45°; a perpetual calendar, table of twelve stars for finding time at night, and a zodiac/calendar scale; a line for finding the hour and azimuth in a particular latitude; a line of natural versed sines which has a corresponding line on the lower edge of the upper arm. Above this is a line of lines, a line of general sines for hour and azimuth at one particular latitude (51°30′) and on the outer edge is a line of 24 hours, 360° and the twelve zodiacal signs for finding the hour by moon and fixed stars. At the end of the arm are tables of week-days, leap years and epacts.

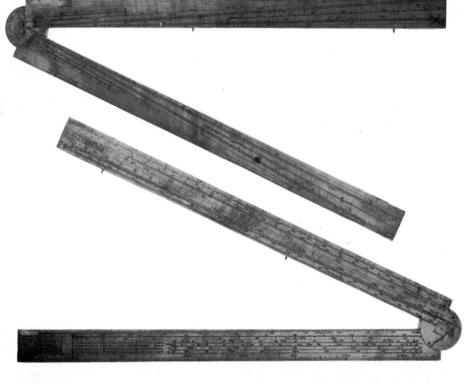

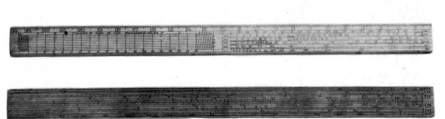

Fig 73
English sliding Gunter rule, top one signed 'W. & S. Jones, 30 Holborn, London', c. 1830, length 789 mm (31 in.), boxwood and brass.

Fig 75 *above*
**English station pointer, signed 'CARY
London',** *c.* 1830, diameter of circle
203 mm (8 in.), brass with wood
extension legs, silver scale.

Circular ring with three pointers.
The scale is divided to 20′ reading by a
vernier to 1′ of arc with a reading
microscope mounted at the centre.

Fig 74 *below*
**English Gunter rule, signed 'JONES
LATE WELLINGTON CROWN COURT SOHO
LONDON',** *c.* 1880, 605 × 45 mm
(23¾ × 1¾ in.), boxwood.

Centre
**English parallel rule, signed
'W. RANSLEY MAKER STANSGATE LONDON'**
ebony with brass struts, 457 mm (18 in.)
long.

Bottom
**English rolling rule, signed
'CRIGHTON LONDON' and with the mark
of the East India Company,** 610 × 76
mm (24 × 3 in.), brass.

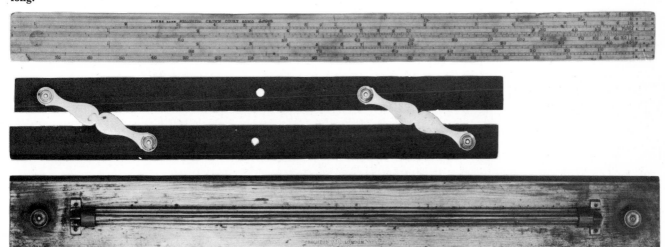

65

comprised of a heavy brass ruler about a foot or more in length, without any calibration. As it was at least three inches wide it served as a parallel ruler, with the added advantage of two knurled wheels on the underside of which it could be smoothly propelled across the chart.

The station pointer was invented by Murdoch Mackenzie in the 1780s, and was a useful device for fixing a position along the coast. It consisted of a circular band of brass, with three adjustable arms which, when laid flat on the chart, could be manipulated so that known bearings could be identified on the chart, and thus the position of the ship calculated (fig. 75).

Weather-forecasting, too, improved from the days of the almanac. A marine barometer was adapted from the stick type at the beginning of the nineteenth century and would be mounted in gimbals on a brass plate attached to the bulkhead. The bottom of the stick was loaded and covered with brass so that the barometer would remain stable in a vertical position in a seaway (fig. 76).

The Determination of latitude

Copernicus and his contemporaries used astronomical tables for determining latitude known as the Alphonsine Tables after Alphonso of Seville for whom they were first arranged by two Jewish scholars, Jehuda ben Moses and Isaac ibn Sid of Toledo, in 1272. They were designed to be used with an astrolabe, but the astrolabe itself could only be used in those latitudes for which plates had been provided, and was difficult to use at sea because of its high wind resistance. The first solar declination tables computed with a calendar for universal use were calculated between 1473 and 1478 by the chief astronomer and cartographer to Manuel the Great of Portugal, the Jewish scholar Zacuto of Salamanca (*c.* 1450–1515), who was expelled from his native Spain by the Inquisition and ultimately sought refuge in Tunis and Turkey. The tables were entitled the *Regiment of the Sun* (implying the rules or regulations of the sun), and were checked off Guinea in 1485. Zacuto and his pupil Joseph Vizinho, a Jew who was forcibly converted to christianity, were consulted by Columbus, and formed part of a nucleus of Jewish scholars who were most active before the sixteenth century in the advancement of science and were persecuted by the Catholic Church. The need for solar tables arose because of the southerly journeys down the west coast of Africa initiated in about 1415 by Prince Henry the Navigator (1394–1460). As the ships pressed southwards, the Pole star, Polaris, dropped lower on the horizon astern, and the nearer they got to the equator, the less they were able to use it as an orientation guide.

The advantage of using solar declination for ascertaining latitude was that it could be used north or south of the equator. When more information regarding the constellations of the southern hemisphere became available, declination tables were prepared for them by Joaõ de Lisboa in 1505. The Southern Cross replaced the North star, with the Cock's Foot as its complement.

Used by Columbus, the *Regiment of the Pole Star* was the instruction book of the mariner, designed for him to find his latitude by the Pole star and its 'Guards', which sometimes known as the two brothers were the two stars nearest to Polaris in the constellation of Ursa Minor. The tables only served as a rough guide, for the elevation of the Pole star changes across the centuries. At the present time it is 1° from the celestial pole, whereas in 1500 it was $3\frac{1}{4}$°, and in 1000 7° from the pole. In classical Greek times the nearest star to the pole was not Polaris but Kochab, the bright star in the constellation Ursa Minor, which later became one of the Guards. This phenomenon, called precession, was known in Columbus's time, but the *Regiment of the Pole Star* took no account of it.

Martin Cortes writing in 1545 referred to the stars of the constellation Ursa Minor (the lesser Bear) relative to the 'head', 'foot', 'right arm' and 'left arm' of an imaginary human figure in the sky with the Pole star at its axis. His head was 'above' the Pole star (i.e. north), his feet 'below' and his arms to right and left. Sailors, having memorized the midnight position of the stars of Ursa Minor for the fortnight (their position shifted anti-clockwise about 15° or one hour during this period), were able to tell the hours before and after midnight from the midnight position of the Guards in relation to the limbs of this imaginary figure. If, for example, in mid-April, it could be said 'midnight in the left arm', then by the end of April, midnight would be 15° or one hour above the left arm. Looking at the stars of Ursa Minor the seaman could calculate the time from their known midnight position.

Shakespeare's audience would have clearly understood his analogy.

> 'The wind-shaked surge, with high and monstrous mane
> Seems to cast water on the burning bear,
> And quench the guards of th'ever-fixed pole.' (*Othello* A. II, i, 13–15)

Rules for using the Guards in whichever position they happened to be were laid down in the thirteenth century by John of Holywood (Sacrobosco) in *De Sphaera Mundi*, a textbook on the sphere, spherical trigonometry and astronomy (fig. 28) and were in use for 300 years.

The Nocturnal

From his position on earth, the observer noted that the Guards were in a different position at the same solar time each night, so that they appeared to move anti-clockwise around the Pole star, and as timekeepers lose one hour in every fifteen days. The reason for the discrepancy between solar and sidereal (astronomical) time is because the sun is constantly moving in relation to the stars.

An instrument entitled a Nocturnal (fig. 79) first mentioned by Martin Cortes in *Arte de Navegar* 1551 was a practical aid for telling the time at night, devised to take this phenomenon into account. To use the words of Andrew Wakeley, a writer on navigation, 'By it may be found the Hour of the night and bearing of the Guards and the Declination of the North Star from the pole' (*The Mariner's Compass Rectified . . .* 1664).

The instrument consisted of two discs, one over the other. The outer disc bore a handle, and was inscribed with a scale divided into twelve points indicating the twelve months. The upper disc was divided into twenty-four equal parts for the twenty-four hours and each hour divided into quarters. These twenty-four hours are denoted by teeth so that they can be told at night, two of which are longer which represented 12 o'clock. A long movable index on the upper plate was fitted by a rivet to the two discs, and in the centre a hole was provided for the observer.

The nocturnal was made to be used with either Ursa Minor or Ursa Major, or with both. When it could be used with either, the handle fixed to the outer disc was movable. To find the time a mariner would turn the upper disc until the longest tooth is against the day of the month on the outer disc, and bringing the instrument to the eye, suspend it by the handle with the plane nearly parallel to the equinoctial, then, observing Polaris through the centre hole, move the index about until he could also see the Guard – Kochab for Ursa Minor or Merak or Dubhe for Ursa Major. The tooth under the edge of the index is at the hour of the night on the edge of the hour circle, which can be counted from the longest one, which is, as we know, midnight.

The Sea quadrant

All elevation-finding instruments have to be used with tables of some kind, from which the observer's position can be calculated. Sophisticated tables for adjustments for parallax, variation and deviation etc., were published later, but with ever-improving altitude-finding instruments, and the mariner's rule of thumb seamanship calculations, latitude determination became increasingly accurate. Tables of all kinds were made available to the navigator by either private- or government-printed publication in all the maritime countries of the world. Only after a slow continuous process

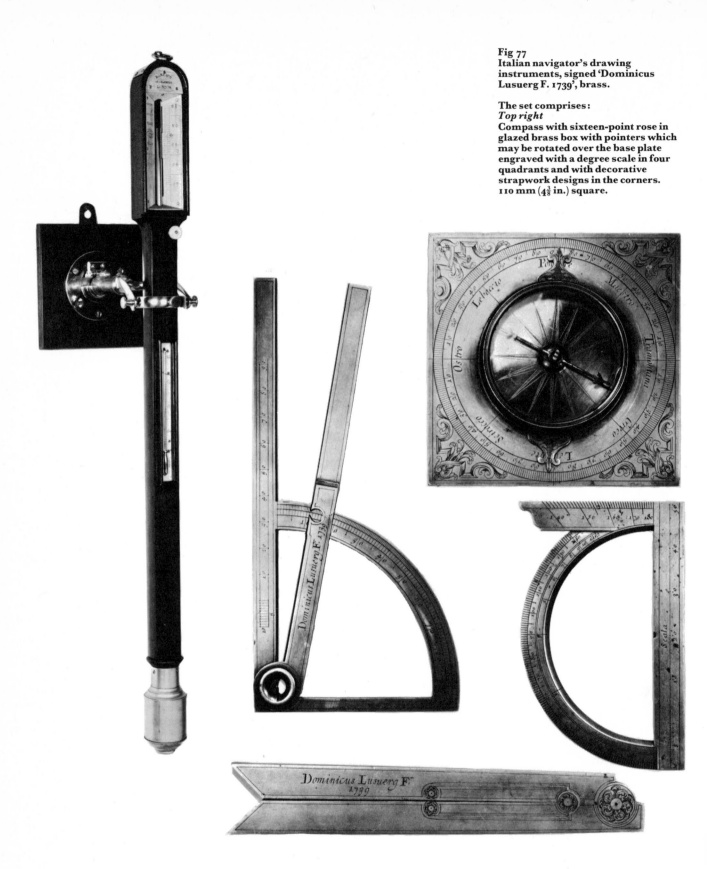

Fig 77
Italian navigator's drawing
instruments, signed 'Dominicus
Lusuerg F. 1739', brass.

The set comprises:
Top right
Compass with sixteen-point rose in
glazed brass box with pointers which
may be rotated over the base plate
engraved with a degree scale in four
quadrants and with decorative
strapwork designs in the corners.
110 mm (4⅜ in.) square.

Fig 76
Marine barometer inscribed
'BARRAUD 41 CORNHILL LONDON', *c.*
1840/50. Overall height 914 mm
(36 in.), mahogany case, with brass
cistern cover and ivory scales, brass
gimbal and bracket. The scale is
divided from 27–31 inches in tenths,
reading against a vernier. A
thermometer is inset in the case.

Right
Protractor engraved on the arc
0–180° and with a right angle arm
140–180°. The horizontal arm is
marked 'Scala' and divided 0–50,
110 mm long (4⅜ in.).

Bottom
Set square (folding) 175 mm long
(6⅞ in.).

Left
Plotting square with adjustable index
pivoted at the right angle. The arc is
engraved with a degree scale 0–90°.

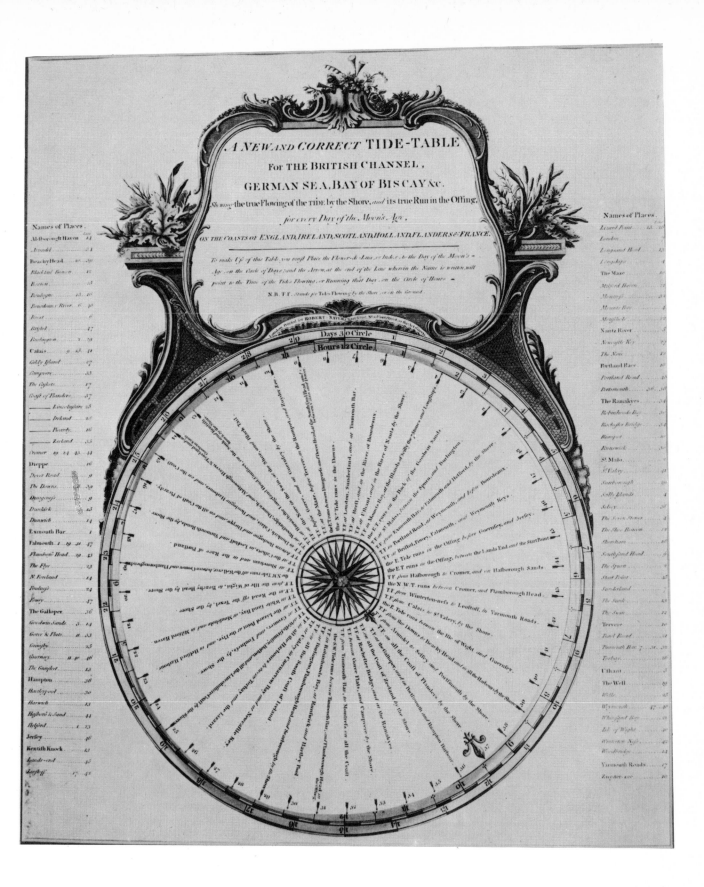

Fig 78
English tide calculator, signed
'London. Printed for ROBERT SAYER,
chartseller, N.º 53 Fleet Street, as the
Act Directs', c. 1770, printed paper
coloured by hand, 504 × 422 mm
(19¾ × 16⅝ in.).

A simple volvelle device which
provided according to its inscription
and directions for use: 'A New and
Correct Tide-table for the British
Channel, German Sea, Bay of Biscay
&c. shewing the true flowing of the
tide by the shore, and its true run in
the offing, for every day of the moon's
age, on the coasts of England, Ireland,
Scotland and Holland, Flanders &
France.

'To make Use of this Table you must
place the Flower-de-Luce, or Index, to
the Day of the Moon's Age, on the
Circle of Days; and the Arrow, at the
end of the Line wherein the Name is
written, will point at the Time of the
Tide's Flowing, or Running that Day,
on the Circle of Hours.'

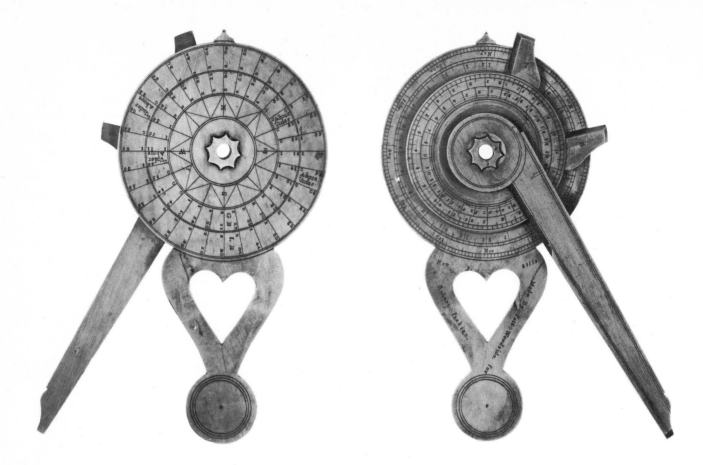

Fig 79
English nocturnal, signed 'Made by
Patt. Woodside for Robert Jackson',
c. 1755–60, length 234 mm (9¼ in.),
diameter 108 mm (4¼ in.), boxwood.

Pierced heart-shaped handle marked
'New stile' presumably referring to
the calendar scale on the edge of the
instrument. On the central disc are
two hour scales (1–12 ×2 and 1–24) and
star shaped brass sighting eye
holding the two plates and the index
in place. On the back is a pole angular
distance table. For use with both
bears.

throughout more than a century of international agreements has the world's astronomical information been shared, to avoid duplicate printing.

Great Britain now produces, to the highest standards of precision, the Ephemerides of the sun, moon and planets, which are then shared with the official almanac-producing agency of every nation in the world. Simplification of the tables is such that what would have taken three hours to calculate in 1800 will nowadays only take an experienced navigator a few minutes.

One of the first elevation-finding instruments was the sea quadrant. Originally the tool of the astronomer and surveyor, this instrument was first used by mariners in the fifteenth century. It was a simple arc of a circle made of boxwood or other close-grained wood or brass, with two sighting pinnules along one straight edge, which were directed towards a celestial body. A plumb bob attached to the apex swings across an arcuate scale graduated 0–90° to show an altitude reading. The device was also used for linear measurement from the departure port.

Before leaving, the navigator took a pole sight when the Guards were in a particular position, and made a note – probably on a slate – of where the plummet cut across the scale of his quadrant. Subsequently during the voyage, he took a sight of the Pole star when the Guards were in the same positions and again noted where the bob cut across his scale. He had been taught that every degree division represented 16⅔ leagues of 3 miles to the league, so he was able to check his distance north or south from his port of departure. This instrument was a practical tool for measuring distance off the coast of Africa, when the navigator could go ashore to take his reading, but the plumb bob instrument was unsatisfactory at sea.

The Cross staff

The cross staff (fig. 80), sometimes known as the fore staff, was first described in 1342 by Levi ben Gerson, a Languedoc Jewish scholar known for his epic work *Milhamot Adonai* (The Works of the Lord), called *The Works against the Lord* by his orthodox contemporaries because of his radical religious opinions. Also called a Jacob's staff, because it resembled the constellation Orion, which was recognized by that name on medieval star maps, it was an instrument for measuring distances between two stars or the angular elevation of a star or the sun above the horizon. About thirty inches long, it was made of approximately half-inch section close-grained wood on which scales were calibrated on all four sides. Three or four alternative cross pieces or 'transoms' could be moved up and down the shaft, one at a time.

To use the cross staff, a navigator would fit one of the transoms, point the staff at the sun, and rest the opposite end of it on the bone beside the eye. The bottom end of the transom should touch the horizon, and the top edge of the lower limb of the sun, by sliding the transom up and down the staff as necessary. The altitude of the sun could then be read off the appropriate scale.

The object of having more than one size of transom was to allow as much manoeuvrability as possible to ensure an accurate reading. The disadvantages were that the user had to spread his angle of vision to see two objects at the same time, and he also had to look directly at the sun. Eventually, a piece of coloured glass was used to shade the glare. The cross staff was also used as a backstaff, in that the observer could turn his back to the sun, fit two transoms on the shaft, the largest and the smallest, and take a reading of where the shadow of one fell on the other.

The instrument was also known as a *balestilha, baculus,* or *arbalista* as it resembled a cross bow, hence the term 'shooting the sun', the expression still used for taking a sun sight (fig. 82).

At first used by astronomers, it was soon taken up by navigators and its use became widespread. Vasco da Gama was shown a cross staff by his Arab pilot from the African coast, now Kenya, when he was taken aboard

Fig 80
English cross staff, signed 'Thomas Tuttell Charing + Londini fecit', *c.* **1700, 786 mm (31 in.), bone or ivory.**

Square section stave, with four cross pieces for 10°, 30°, 60° and 90°. The 10° cross may be used for back observations of the sun. With incised floral and geometrical decorations, original lyre case.

Part of a set of navigation instruments probably made for a presentation.

Fig 81
English trade token, signed 'DIXY PAGE
AT YE ANCHOR AND/MARRIN[ER] IN EAST
SMITHFIELD/HIS HALFF PENY 1667'

The obverse shows a mariner sighting
with a cross staff and standing beside
an anchor.

Dixy Page (*fl.*1667–74) was a bookseller
specializing in mathematical and
navigational works.

Fig 82
The cross staff in use from Nathaniel
Colson *The Mariner's New Kalendar
...also the description and use of the
sea-quadrant, fore staff and
nocturnal...* London 1722, p. 64. First
published in 1675, Colson's work went
through numerous editions before
being revised in 1761 and 1764 by
William Mountaine.

to guide the explorer to India in 1497–9. Cross staves were still being used in Holland until the nineteenth century and sea conditions can weather a spar very rapidly into an antique state, so, should a collector be fortunate enough to find one, this fact should be borne in mind. However, cross staves of whatever vintage, especially when complete, are a joyful addition to any collection.

The Mariner's astrolabe
The mariner's astrolabe evolved from large wooden circles with alidades suspended from frames used originally for astronomy; an account of one of these exists describing how it was used to observe the solar eclipse of 3 March 1337.

The occasional reference to an astrolabe either made of wood, or made of metal with spokes, distinguishes the mariner's astrolabe from the many references to its much older namesake. Amongst early references the most famous, especially as it is illustrated, is that of Pedro de Medina in 1552 in *Regimiento de Navegacion* published in Seville in 1552. The first English author to depict it was probably William Bourne in 1567 in his *Almanacke and Prognostication for iii yeres, with serten Rules of Navigation*, but as a copy has not survived, it is only known by his next editions for 1571, 72 and 73, although it is presumed that the same wood-cut was used previously.

The planispheric astrolabe was introduced into Europe by the Arabs and the earliest known survivors date from the tenth century. It was too light to be used at sea, although it was carried to sea but used after landfall. One wonders why it took so long for an instrument adapted from it to be made available to the navigator. R. C. W. Anderson has quoted several sources (in his *The Mariner's Astrolabe* Edinburgh 1972) which suggest that astronomical circles were too conservative, or that there was a lack of imagination on the part of teachers of astronomy to consider the problems of the humble seaman. It was probably due to lack of incentive.

During the Middle Ages shipping in the northern countries consisted of small traders, coasting from port to port. There was insufficient shipping to transport the Crusaders, who travelled overland to Mediterranean ports before boarding ship for Palestine. Mediterranean mariners were the heirs to a long history of maritime activity, and, being used to the old ways, were slow to develop new methods.

While the astrolabe perfected by the Arabs was available to astronomers, a certain class distinction prevented its infiltration to the mariner. Wealthy ship-masters, or even a passenger, might well possess such an instrument, but by its construction it was limited to use on land. There seems to have been no stimulus to adapt what was, after all, a highly sophisticated altitude- and time-finding instrument to the needs of the mariner. That was until the advent of Prince Henry the Navigator. His plans for the discovery of the west coast of Africa and what lay beyond caused considerable interest in maritime matters. When his pilots sailed due south they needed new instruments, for those with which they had been provided were inadequate, so all available mathematicians were assembled to improve the situation. The rediscovery of the Azores in 1432 and the exploit of Gil Eannes in 1434 when he rounded Cape Bojador well out to sea continued to emphasize the need for more information, for the riches of the Orient were awaiting the intrepid mariner who could find an alternative route to the Molluccas. There at last was the incentive, and in the pressure exerted by the demand a break-through was accomplished. Necessity being the mother of invention – or as G. K. Chesterton put it, 'Progress is the mother of problems' – the mariner's astrolabe appeared towards the end of the fifteenth century although no definite attribution to its invention can be made. Both Christopher Columbus and Sir Francis Drake were familiar with this instrument, but few have survived, most of which have been recovered from wrecks quite recently. A classification scheme has been devised by D. W. Waters, based on the distribution of additional cast metal retained as ballast to increase the weight of the instrument so as to make it more stable in adverse conditions on board ship.

Fig 83
**Remains of a mariner's astrolabe,
not signed,** *c.* **1600, diameter 179 mm
(7 in.), brass.**

**Wheel-type mariner's astrolabe with
base ballast weighing 4lb. 8oz. The
instrument was originally divided
0°–90°–0° round the upper quadrants
with stars dividing the degree marks.
A fragment of the alidade survives.
The instrument was found at Lyme
Bay, Dorset in 1967.**

Fig 84
**Portuguese (?) mariner's astrolabe,
signed, 1555, diameter 222 mm (8¾ in.),
brass.**

**The earliest surviving dated
mariner's astrolabe, wheel-type with
base ballast, weighing 6lb. 6oz. The
scale is graduated for zenith
distances (90°–0°–90°). A mark made
up of five circles has been
interpreted by Destombes (***Revue
d'Histoire des Sciences*** xxii, 33, 1969)
as the sign of Lopo Homem, a 'master'
of nautical instruments in Portugal
1517–1565. On the reverse the name
'ANDREW SMYTON 1688' is stamped.
Andrew Smyton (Smieton) was a
Dundee shipmaster. The instrument
was subsequently owned by the Rev
Dr Thomas Dicks (1774–1857).**

In appearance the mariner's astrolabe resembles a four-spoked, cast bronze wheel, with a jointed ring at the top for the navigator's thumb and a pivoted alidade with slits through which he peered at the celestial body to check altitude. Calibrated 90–0–90 across the two top quadrants for zenith distances, usually associated with Portuguese manufacture or marked for altitude heights suggesting another school of makers, they vary considerably in diameter and weight, but an average size would appear to be between 13 and 18 centimetres (5–7 in) in diameter and weighing 2 to 3 kilos (4–7 lb.) (figs. 84 and 83). A difficult instrument to use, it was nevertheless progressive, and of considerable advantage to Portuguese and Spanish pioneer explorers. The British adopted the instrument although the only one extant of definite British origin is signed by Elias Allen dated 1616. Some of those recovered from wrecks around the coast of Ireland were once aboard the Spanish Armada fleet.

The Backstaff

The blinding glare of the sun in an observer's eye when he took his noon sight caused him to turn his back with his cross staff, and it seems natural enough now that a special instrument was evolved to perform this function accurately. Sailor-scientist John Davis (c. 1550–1605), sea captain and explorer, who gave his name to the Davis Straits and whose voyages around the coasts of Canada are carefully delineated on old globes and maps, wrote two books: *The Seaman's Secret* in 1594 and *The World's Hydrographical Description* in 1595. In the second of these he advocated the use of the terrestrial globe as an instrument of practical navigation and described sailing on a rhumb line; in the first he made an important contribution to navigation by describing his backstaff.

This was an ingenious improvement on the quadrant, cross staff and mariner's astrolabe for taking an elevation. Usually made of wood section $\frac{5}{8}$ inch by $\frac{3}{4}$ inch (1·6 × 2 cm.) in lignum vitae, the instrument was formed with a main limb about 24 inches (61 cm.) long, with a right angled accessory termed a horizon slit on the end. The quadrant was divided into two arcs made of boxwood. The smaller arc, calibrated to 60°, was mounted

Fig 85
English backstaff, inscribed 'JAMES RICH 1771' but probably late seventeenth or early eighteenth century, length 580 mm (22⅞ in.), mahogany and boxwood. Firmly braced to the central staff of the instrument are two arcuate arms. The smaller of these – the shadow vane arc – is divided into 60°, the 10° intervals being marked along the upper edge, the 5° intervals along the lower. The 30° mark having originally been engraved twice by mistake on the scale, the values 40°, 50°, and 60°; 45° and 55° have been engraved over the incorrect values (10° lower) underneath. The larger arc – sight – vane arc – of the instrument is divided into 30° reading to 6′ by a diagonal scale. With *fleur de lys* and Tudor rose decorations. The sighting vane, horizon slit and shadow vane are all missing. For altitude readings only.

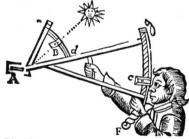

Fig 86
The backstaff in use from Nathaniel Colson, *The Mariner's New Kalendar … also the description and use of the sea-quadrant fore-staff and nocturnal*, London 1722, p. 63.

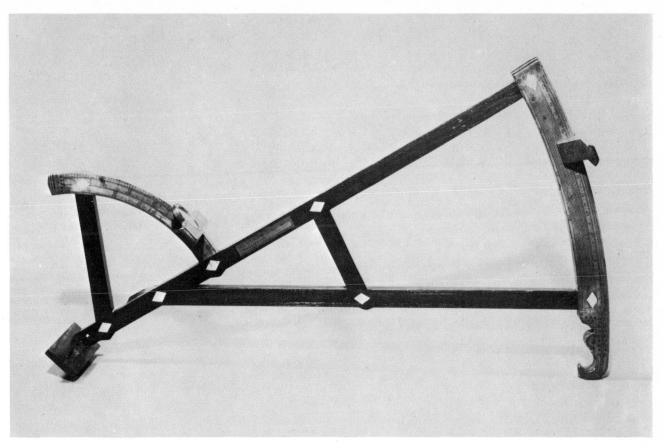

Fig 87
English backstaff, signed 'Made by
I. GOATER near Union Stairs in Wapping
LONDON', *c.* 1760, overall length 631 mm
(24⅜ in.). Lignum and boxwood with
ivory diamond insets at the joints.
The arcs are decorated with *fleur de
lys,* Tudor roses, stars and suns. The
larger (sight-vane) arc is numbered
from 0–25° along its lower edge and
65–90° along the upper. A common set
of diagonals enables readings to be
taken to 10′ of arc. Along the outer
limbs (i.e. the 0–25° scale) a further
sub-division makes possible readings
to 5′ of arc. The smaller arc (for the
shadow vane) is divided into 60°, the
10° intervals being marked along the
lower edge, the 5° intervals along the
outer. With original spring-loaded
shadow and sight vanes, the latter
with a single lens. The horizon slit is a
modern replacement.

Fig 88
English Hadley quadrant, signed
'GOATER at N° 141 Wapping LONDON', *c.*
1770, overall length of arm 510 mm
(20 in.). Mahogany frame inlaid
boxwood scale reading to 20′ of arc,
two inset ivory plaques (blank) and
brass fittings. Double arm frame with
arched supporting strut and a
movable arm pivoted at the centre of
the circle of which the limb forms an
eighth part (radius is 7 mm or 18 in.).
Exactly above the pivot is a
silvered mirror (index glass) used in
conjunction with a second mirror
(horizon glass) on one of the fixed
arms and an eyepiece on the other for
measuring the altitude of celestial
bodies. The instrument is fitted with a
second mirror for back observation
and two sets of coloured glass shades.
There were several instrument-
makers named Goater working in
Wapping in the late eighteenth
century of whom John Goater was the
senior. The quadrant was possibly
made by Henry Goater, his son or
nephew.

above the main limb and fitted with a shadow vane. The larger arc, calibrated to 30°, was mounted at the far end of the main limb and was fitted with the sighting vane. This arc was further strengthened by the support of an auxiliary limb about 18 inches long (45 cm) fitted to the main limb, and a cross member between the two provided extra rigidity and served as a hand grip.

To use the instrument the observer stood with his back to the sun, and adjusted the shadow vane to within 10° or 15° less than the possible altitude of the sun. He then held the backstaff by the cross member with his left hand, and peered through the horizon slit raising or lowering the instrument until he could see the horizon through the sight vane. He then adjusted the shadow vane until the shadow of its upper edge fell on the upper edge of the horizon slit. The sum of the readings on the two arcs cut by the edges of the vanes would then have been the sun's altitude.

The success of the backstaff was phenomenal. Made almost exclusively in England, although Irish and American ones are known, it was used all over the world for nearly 200 years. Termed the Davis quadrant, the English quadrant or *Quartier l'Anglois (sic)* or *Quartier de Davis* by the French it was the seaman's trusted servant. However, a notable exception to its widespread popularity was in Holland, where the conservative Dutch clung to the use of the cross staff well into the nineteenth century.

It is tempting to think that when improved methods are proposed they are immediately adopted. We have seen how reactionary the seamen were, and how ill-educated. They learned their navigation parrot fashion, without understanding the astronomy or mathematics except in the most rudimentary form. After all, it is still not considered necessary to be a scientist to become a seaman. Only as late as 1920 did sextants become navy issue for the first time in British history, so until then the navigator bought his tools when he got his ticket and with luck they lasted throughout his working life. If the tools served him well, he saw no need to replace them, and if he had bought them second-hand, their life-span was extended. A. J. Hughes reckoned in *The Book of the Sextant* (London 1938) that a sextant made from the early nineteenth century onwards would give fifty years of accurate service.

The social background of the tradesman and his tools has to be considered when looking at the overlap of instruments used apparently anachronistically.

The Hadley quadrant

The American contribution to navigation is too important to pass unmentioned. The most important written works were the first geographical and political manual with navigational guides from port to port by Martin Fernandez de Enciso, published in Seville 1519 (translated into English by Roger Barlow in 1541), valuable sixteenth- and seventeenth-century manuals in Spanish translated into English, in particular *Instrucion Nautica* by Diego Garcia di Palacio published in 1587 in Mexico City (revised, elaborated and restated by Nathaniel Bowdich in 1802), and the first English American work of Benjamin Hubbard. Published in *Charls Towne in New England*. Hubbard's *Paradoxall Chart* a method of great-circle sailing based on Mercator's projection was an important step towards the final acceptance of Mercator's system.

The pressing commercial need for more accurate navigational methods, which is reflected in the offer of a £20,000 prize for a practical solution to the problem of how to determine longitude in Britain, and similar prizes in other countries, produced a ferment of activity in all aspects of navigational science. Working independently and in complete ignorance of each other John Hadley v.p.r.s. (1682–1744) in London and Thomas Godfrey, glazier and natural mathematician in Philadelphia, simultaneously devised an improved form of altitude-measuring instrument which worked on the same principles. The Royal Society recognized the equality of the two and awarded each a prize of £200. Godfrey received his in household furniture.

Fig 89
English Hadley quadrant, signed
'R. RUST, LONDON.' *c.* 1760–70, mahogany
with inlaid boxwood scale and brass
fittings.

The standard mid-eighteenth-century
form of Hadley quadrant. The scale is
divided to 20' of arc.

Fig 90
English Hadley quadrant, not signed,
c. 1780, radius 452 mm (17¾ in.),
mahogany with inset ivory scale,
brass fittings and inset silver plaque.

A standard pattern Hadley quadrant
with brass casing on the lower half of
the index arm. The scale is divided
0°–90° by 20' divisions; vernier with
ivory scale and clamping screw.
Engraved on the plaque is the legend,
'Presented by George McConnell to
Hamilton Cairns as a Mark of Esteem
for His Gallant and prudent Conduct
for staying by the ship Pomona when
diserted by the Captain and all the
Crew except 3 Boys, & bringing her
safe to Har'.'

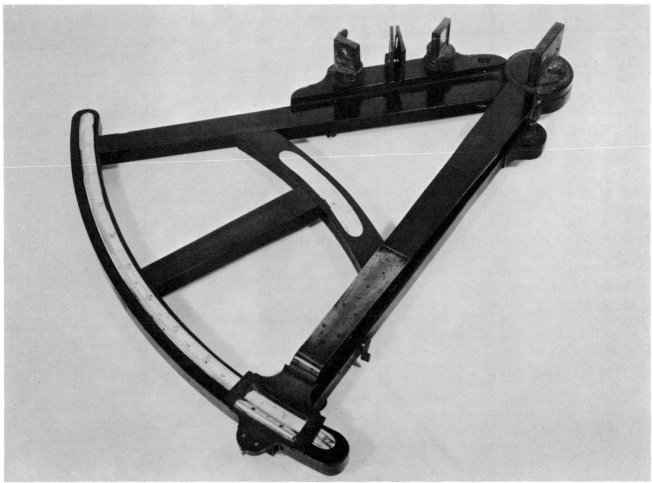

The Hadley quadrant, as it came to be called, although it was later known as an octant, was a brilliantly simple instrument based on the application of optics. Returning to the conception of the sea quadrant with a single arc of 90°, the instrument was made of a triangular frame of wood, lignum vitae or mahogany for strength, with a movable index arm pivoted from the apex. A mirror was fixed at this point that would move with the index arm. An observer would peer through the sighting pinnule placed on the limb, and tilt the instrument until he could see the horizon in the clear half of the second glass fixed on the opposite limb. He then adjusted the index arm until the celestial body appeared to be reflected onto the horizon. Finally he checked the vernier which is fitted at the end of the index arm over the arcuate scale for finite adjustment, and took the reading on the scale.

There are two main principles involved. First, the angle of incidence equals the angle of reflection in a plane which contains the normal to the reflecting surface at the point of reflection. Second, if a ray of light suffers two successive reflections in the same plane, by two plane mirrors, the angle between the first and last direction of the ray is twice the angle between the mirrors. Because the angle between the two mirrors is half the altitude of the object observed, when the mirror on the index arm moves from the parallel through to the angle, *double* the angle will be read on the arc. Thus the arc will read up to 90° although in itself it is only an eighth of a circle (45°), hence the term octant.

The instrument was made in several qualities with or without refinements. Some of the variations available can be assessed from the catalogue dated 1823 of 'Optical, Mathematical and Philosophical Instruments' made by W. & S. Jones:

HADLEY'S QUADRANTS, mahogany, the divisions on wood	2.	2.	0
Ditto mahogany with ivory arch and nonius (vernier) double observation	2.12.		6
Ditto, ebony and brass, best glasses, engine divided &c.	3.	3.	0
Ditto, with tangent and adjusting screws, &c.	3.18.		0
Ebony and brass mounted best sextants, from £4. 4s. to	8.18.		6
A ten inch common brass sextant	9.	9.	0

Metal 8, 9 or 10 inch ditto, framed on a principle the least liable to expand or strain, with adjusting screws, telescopes, and other auxiliary apparatus, divided to 30″, 15″ or 10″ the best for taking distances accurately, to determine the longitude at sea, &c. from 13l. 13s. to . . . . . . 16.16. 0.

The vernier was added for fiducial finite adjustment because of the work of Sissons and Bird on mural astronomical quadrants. A telescope replaced the sighting pinnule, or was offered as an alternative. The index arm was either handsomely engraved, bore a fin for extra rigidity, or was simply plain. Generally speaking the polished wood instrument with a transverse boxwood scale was made up to 1750–60 when brass was introduced for the index arm, either for just the vernier end or for its entirety (figs. 89 and 88). Earlier instruments had the finite scale on the vernier calibrated 10–0–10, while those of the second half of the eighteenth century onwards were calibrated 0–20.

The wooden limbs were blackened or 'ebonized' to reduce glare, and for greater clarity the scales were engraved on ivory (fig. 90). A nameplate was provided for the owner as well as a pencil secreted in the cross member to be used to record data on a small ivory plaque on the back. The instruments were fitted with two sets of coloured glass shades for use with sun, and the earlier type of instrument had a second pinnule fitted on the opposite limb, so that when the horizon below the sun was ill-defined the opposite horizon could be used. This was of greater use at anchor off unexplored coasts where latitude had to be determined.

Fig 91 *Far right*
English octant, signed 'JONES, GRAY & KEEN, STRAND LIVERPOOL', c. 1820, radius 329 mm (13 in.), ebony frame with brass arm and fittings inlaid ivory scale and name plate.

Although following the traditional design of frame, the instrument is fitted with a vernier scale with tangent and clamping screws. The little-used back observation mirror has been abandoned.

Right
English octant, signed 'J. COX LONDON', c. 1840, radius 304 mm (12 in.), ebony frame with brass arm and fittings, inlaid ivory scale and name plate.

Double T frame design with handle, ribbed brass arm with vernier clamping and tangent screws and reading microscope. Plain eyepiece in which the diaphragm may be slid away and a sighting telescope screwed into place. Scale divided 0°–105° reading to 20′ of arc.

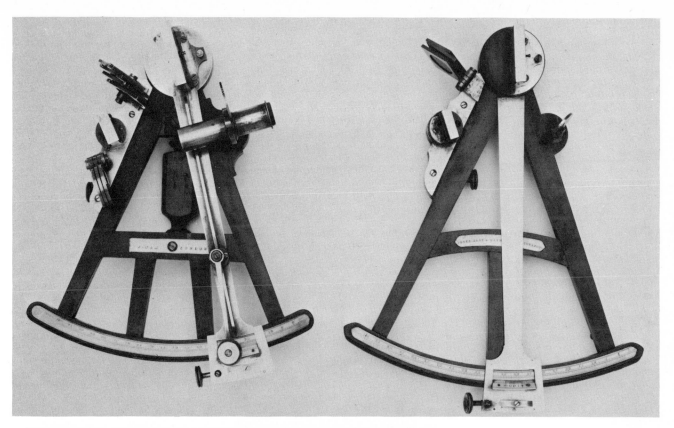

Fig 92
Portrait of Jesse Ramsden
(1734–1800), 1791 Mezzotint, 443 × 356
mm (17½ × 14 in.), published by
Molteno, Colnaghi & Co. Painted by
R. Home, engraved by J. Jones.

Nearly whole length portrait of
Ramsden seated beside his dividing
engine facing left, and with a pair of
dividers in his hand. In the
background is his large wheel transit
instrument.

The size of the instrument was controlled by the fact that the scales on the arc had to be calibrated by hand. When Jessie Ramsden (1735-1800) invented his dividing machine in 1771, this operation could be swiftly, accurately and economically carried out in a smaller area, so the size of the instruments shrank from a radius of approximately 45 centimetres (18 in.) to one of approximately 20 centimetres (8 in.). In this compact form the octant was in use until the end of the nineteenth century. They were cheap to buy (they were offered in a catalogue at thirty shillings) and were less vulnerable than the brass instruments that ultimately replaced them, when used on board a small freighter or fishing boat.

The handsome quadrant or octant was well established as a practical 'no nonsense' tool by about 1790. Perhaps the onslaught of the Napoleonic wars and the enormous demand for instruments for the hastily impressed battle fleet, plus the new mass-produced scales caused this decline in aesthetics, but no nautical instrument had ever had a wider appeal. Wherever there was a need for maritime victualling, octants were sold, frequently with the chandler's trade label in the box, or on the instrument. Octants were made first in England, America and Ireland, and then in

Fig 93
English sextant, signed 'Ramsden London 1277', c. 1794. Radius 220 mm (8⅝ in.), brass frame with wood handle and platinum scale. Double frame with three coloured glass shades for the index glass and three for the horizon glass. Three sighting telescopes. The scale is divided 0°–130° and has a vernier with tangent screw with spiral screw adjustment.

France, the first by Pierre Lemaire (*c.* 1739-60). Ramsden's dividing machine (fig. 92) was made by other manufacturers, for he had received an award for its invention, and held no patents. Spencer Browning and Rust (*fl.* 1787–1842) used one of these to great effect for they must have made scales for nearly everyone.

The initials s.b.r. will be found in the centre of the ivory scale of many octants bearing either another maker's name, a chandler's name or no name at all.

The Sextant

The familiar brass sextant was based on the same principles as the Hadley quadrant, but was intended as an improvement in that the wooden frames of the octant were inclined to distort in humid conditions and caused errors. Also the larger arc of the sextant was more useful.

It is not clear who first made a brass instrument, but Edward Troughton (1753–1836), a founder member of the Royal Astronomical Society, patented a form of brass sextant in 1788. The limb was formed of strips of plate, in duplicate, the two joined together with turned brass pillars. This

Fig 94
English sextant, signed 'Ramsden London', 1390, *c.* 1798, radius 250 mm (9⅛ in.), brass with lignum handle and platinum scale. Bridge-type double frame with four shades for the index glass and for the horizon glass; two sighting telescopes. The scale is divided 0°–130° in minute divisions and has a vernier with tangent and clamping screws with a reading microscope.

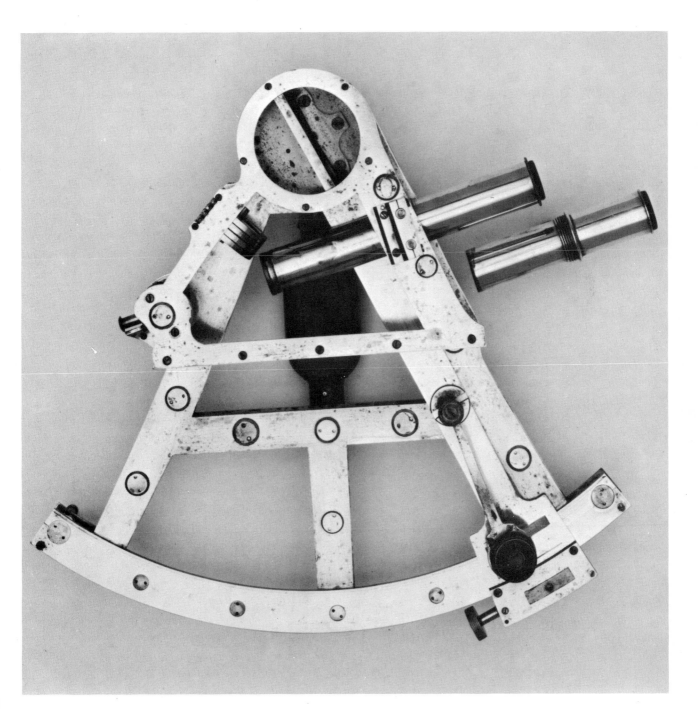

type of sextant, termed the Troughton type, or double frame, was being made as late as 1830. In the meantime, Jesse Ramsden and others were experimenting with thicker gauge metals and other forms of manufacture to ensure rigidity. The so-called 'bridge type' made by Ramsden (fig. 93) with the serial number 1277, was made before that shown in fig. 94, which uses Troughton's pattern and is numbered 1390. Another bridge type, that made by Fraser (fig. 95), was probably contemporary with Ramsden's. Of all the experimental types, the Troughton seems to have been the most popular.

Makers really overcame the problems when they began to use alloys which could take strain. Bell metal alloy containing 7 per cent tin became standard in the nineteenth century, and an optimum weight of four pounds was deemed the most serviceable. Curiously enough, the most difficult part of the instrument to make was the plane mirrors. The two faces had to be ground parallel and silvered in the old way with mercury

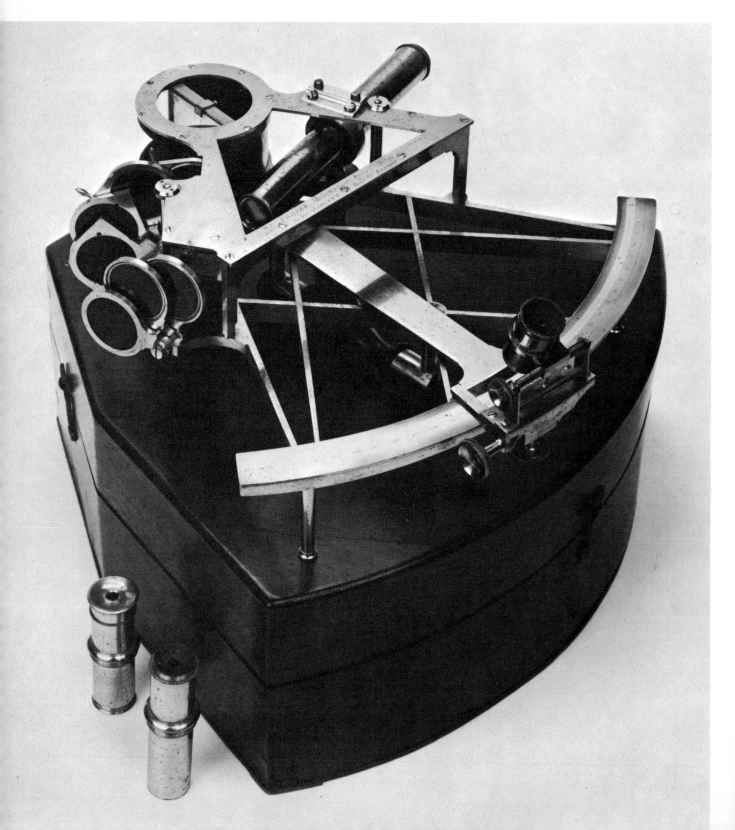

The National Physical Laboratory.
TEDDINGTON.

THIS IS TO CERTIFY THAT SEXTANT No. 3823

D. Shackman & Sons, London & Chesham

Fig. 96.
English sextant, signed 'MADE IN ENGLAND BY D. SHACKMAN & SONS, LONDON & CHESHAM', no 3823, *c.* 1940, brass and aluminium in mahogany box.

A typical twentieth-century sextant with certificate from the National Physical Laboratory, Teddington, illustrating the simplification, but lack of basic change, in the design of the instrument.

and tinfoil. The problem lay in the grinding of the glass, for if it was not absolutely flat the instrument would be inaccurate.

Telescopes provided with the sextant were developed with great care. The eyepieces are either of the Huygens or the Ramsden type, which work on different optical principles, although the results are almost the same. The main difference is that the Ramsden can be used with crosswires for fiducial measurements whereas the Huygens may have crosswires, but only so as to mark the centre.

The object glass is achromatic to avoid spherical and chromatic aberrations which the four spherical surfaces of the meniscus or concave flint glass lens and the bi-convex crown glass lens overcome when bonded together.

When we confidently read of the development of the sextant, it may come as a surprise that as recently as World War II a serious problem confronted both the American Maritime Commission and the Ministry of

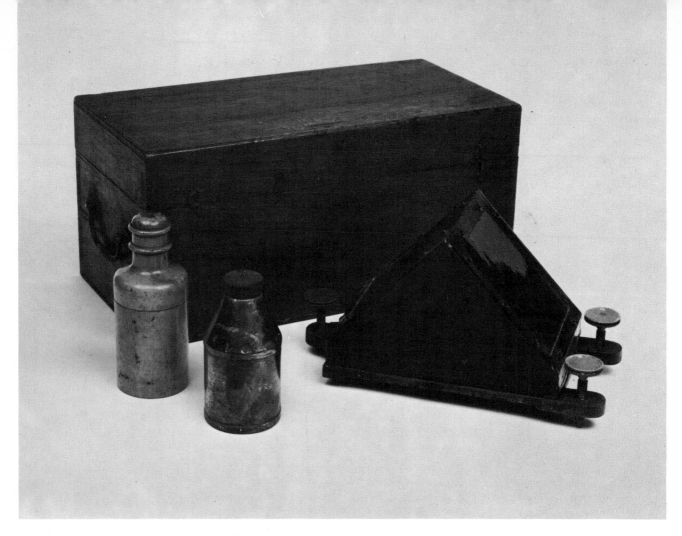

Fig. 97.
Artificial horizon, not signed,
nineteenth century. Size of box
335 × 143 × 135 mm (13¼ × 5½ × 5¼ in.).
Triangular prism with three levelling
screws and spirit level. Two mercury
bottles of brass and boxwood
respectively. In original oak case. The
instrument was used for obtaining
accurate altitude readings on shore
when high precision was required
(i.e. for setting chronometers). When
the artificial horizon is properly
levelled and a reflection of the sun
obtained, the angle between the sun
and its reflection is measured, and
then halved to find the altitude.

Supply in connection with sextants. The loss of shipping in the Atlantic at
the rate of many million tons a year led to an all-out drive to replace lost
equipment. By 1943, because the industry was overloaded, urgent requests
were sent to other firms, experienced in precision work, both optical and
dividing, capable of making sextants to within the strict limits imposed by
the National Physical Laboratories. One of those that responded was
David Shackman & Sons of Chesham, Bucks., whose wartime contracts
including gunsighting equipment and pre-war craftsmanship of fine
jewellery made them eligible. Reuben Shackman, awarded the M.B.E.
for his contribution, tackled the many problems involved. He persuaded
the Science Museum in London to loan him an old Ramsden dividing
machine found in their cellars, which he had semi-automated by rigging
it with an electric motor, while retaining all its other characteristics. A
place was made for the sextant at the centre with a toothed form cutter
mounted on a rotating spindle for 'gashing' the worm wheel teeth.
Ramsden's machine provided the least of the problems – the worst were

the optical ones, which were resolved by new processes using Rhodium, a metal which is impervious to sea water. Novel features were introduced which resulted in better production by unskilled labour, which comprised mainly women supervised by a craftsman (fig. 96).

The Determination of longitude

The fixing of latitude had never presented as great a problem to the navigator as the determination of longitude. This is complicated by the fact that it is measured in an east-west direction from an arbitrary datum point. Travelling westwards, the sun is at noon at each meridian in turn, until the complete circle of 360° of longitude is completed, covering 15° of longitude in one hour. If at the moment of the sun's meridian at his position the observer knew the time at the datum point, he could calculate the difference in time, and so the meridian of longitude.

The relationship of time with the determination of longitude was mentioned by Gemma Frisius in 1530, who suggested using a clock for its calculation. However, no portable clock was sufficiently reliable, and none suitable for use at sea. Until such a timepiece was perfected navigators relied on other methods, notably computation of time by sandglasses and of distance by the log.

Sandglass

Distance was an abstract hypothesis until the accurate measurement of the earth was accomplished, which is still in dispute. Through centuries of supposition, the minute mile evolved, that is to say a minute of *arc* – not a minute of time (which could be confused with athletics) – from the premise that 60 *nautical* miles (6080 feet) equal 1° of arc. Each degree is divided into 60 minutes and each minute is divided into 60 seconds, of arc, indicated by " for minutes and ' for seconds. Relating distance to time meant the use of the sandglass as a time piece. A sandglass, or running glass, contains either sand, marble dust, alum, iron filings or powdered eggshells. Consisting of two hand blown bulbs with flanged ends, sealed with putty or wax and covered with linen and twine, they were made in many sizes and varieties for all purposes for the measurement of intervals of time before clocks and watches were reliable. Amongst those used at sea was the half-hour glass, which was turned by the ship's boy, who rang a bell to mark the event. At eight bells the watch was changed, and the series of eight bells was recommenced, starting at noon. The system of the four-hour watch and the two dog watches was first described in English by Thomas Hariot. Because of human error the turning of the half-hour glass was in need of constant correction, and the glass was restarted by being turned upside down every few days at the instant when a pin stuck into the binnacle threw its shadow onto the compass fly at due north.

For accuracy several glasses were carried and they were sometimes made four to a case, either all for the same length of time, or for the hour, ¾ hour, ½ hour and ¼ hour. Ships' glasses were enclosed in stout wooden frames with four to eight legs, and the four-hour or watch glass was provided with two cords (beckets) which were passed through the base plates and rove and spliced to form double bights from which the glass could be suspended next to the binnacle.

Log, line and patent logs

Longitude was a baffling problem. William Bourne (*fl.* 1560–97) advised his readers not to worry about it too much but to rely on dead reckoning for determining an approximate position: 'Dead', from deduced, reckoning was used from a record of the ship's continuously changing speed and course.

According to Pedro de Medina 'the most a ship advances in an hour is four millas and with a feeble breeze three or only two. . .'. Columbus used the Roman or Italian 'milla' which was equal to 4,842 feet or 0.7936 nautical miles. Four millas made a 'legua' or 3.18 nautical miles, or 3.67 miles. The English used the league, three miles, and the fathom, six feet,

Fig. 98.

Top
English (?) sandglass, not signed, *c.*
1700. Height 195 mm (7¾ in.), turned-wood stand with four shaped pillars,
duration 30 minutes.

Left
French sandglass set, not signed,
eighteenth century, 170 × 160 × 75 mm
(6¾ × 6¼ × 3 in.). Engraved brass
stand with ten pillars and
architectural frieze. Duration 30
minutes and 60 minutes.

Right
French sandglass, not signed,
eighteenth century with divided
ampoule, brass frame with six
pillars, duration 20 minutes.

Front
Italian (?) sandglass set, *c.* 1700.
Fitted wood stand, duration 10, 20 and
30 minutes.

taken originally from the distance between the tips of the fingers when a man's arms are fully outstretched.

'That in order to know the course of the ship he must set down in his register how much distance he has made according to hours' de Medina instructed. An unsuccessful attempt to measure distance at sea had been made by Vitruvius (14 B.C.), a Roman architect and engineer who described an instrument in detail called a 'way measurer' or 'hodometer' for determining distances covered by a vessel. It comprised a paddle wheel about 120 centimetres (4 ft) in diameter attached to the hull, which actuated a train of three gears as the paddles dipped into the water. The last gear operated a gate which caused a small round pebble to drop into a box at each revolution of the gear. By counting the pebbles in the box at the end of a two-hour run measured by the half-hour glass, the rate of motion might be calculated (described in the *Saverien Alexandre Marine Dictionary* 1769).

De Medina wrote, 'The distance sailed is estimated by the eye by established principles' (observing the run of water along the hull). The phenomenon that a piece of wood thrown into the water appears to remain stationary (disproved by Leonardo da Vinci) led to the use of the common log as a measurement of dead reckoning. The name 'log' appears to be of Saxon origin 'a hewn trunk or branch of a tree' but it has also been suggested that the log was so named from the Dutch 'liggen' to lie down, and it now seems natural enough that the name log should come to be applied to this primitive contrivance. The method was this: a piece of wood attached to a line was flung overboard astern, and the line played out for a given period of time. The line was then brought in and measured so that the distance covered in the period and therefore the speed of the vessel could be calculated.

William Bourne *fl.* 1565–97 (the modern spelling is mine) wrote that 'dead reckoning is an uncertain guess, and if you please to call it probable conjecture, you shall grace it with the uttermost'. Great store was not placed on its accuracy, in spite of its popular use. Some were better at it than others; Columbus was remarkably qualified as his journal testifies. Edmund Gunter (1581–1626), Professor of Astronomy at Gresham College, first described the Dutchman's log, which was a chip of wood thrown overboard at the bow, and the time measured between two marks on the gunwhale (or two seamen strategically placed). The rule was commonly summed up as:

'As the time given is to an Hour
So the way made, to an hour's way'

The seaman, armed with Gunter's scale, transcribed fathoms or feet per minute or half minute (using one or two turns of the half-minute glass) into miles per hour, but an even simpler mechanical method evolved. The log-line came to be knotted at regular intervals. The intervals between the knots were directly proportionate to miles, so that speed could be calculated by the number of knots fed out in a given time.

Such was the seaman's dislike of change that the log-line continued to be knotted at 7 fathoms with a 28-second log glass in spite of a more rational 51 feet per half minute proposed by Richard Norwood based on his accurate geodetic measurements. When the Board of Longitude was formed in 1714, with its promise of prize money, all things navigational became of paramount interest. A series of mechanical logs were patented, the first of which was called the 'Saumarez Marine Surveyor', patented in 1715, after the name of the inventor Henry de Saumarez of Guernsey. Sir Isaac Newton, among others, decried the invention for he had seen another invented by Thomas Savory and neither was satisfactory. In spite of considerable activity, Saumarez received no award.

Several other devices followed, but an invention by a Deptford mariner named Foxon called a hydrometer caused enough interest to be ordered by the Admiralty for Constantine Phipps on his voyage towards the

Fig. 99. *left and right*
English Log, signed 'Foxons Hydrometer By the Kings PATENT, No. 43', *c.* 1772, 213 × 224 × 274 mm (8⅜ × 8¾ × 10¾ in.), brass, steel and oak.

Mechanical log patented by William Foxon, carpenter of Deptford, Kent, December 5th, 1772 (British Patent no. 1028). A spiral log was trailed in the water behind a ship at the end of a line (of maximum length 15 fathoms) which was connected to the mechanism. As the log revolved it twisted the line thus imparting motion to the central arbor of the mechanism. This arbor carries an endless screw and also has a fly wheel to equalize its motion. The endless screw takes into the first wheel which drives a hand round the dial plate divided into twelve knots by divisions of a seventh of a knot. The first wheel also meshes with a second endless screw which connects with a second wheel which drives a hand round a second dial divided into 12 miles by one tenth mile divisions. The pointer of the third dial is geared to turn once for every twenty-four revolutions of the mile dial, i.e. every 288 miles.

Arctic and for James Cook on his second voyage. Both captains found it to be almost accurate in calm weather but useless in heavy seas. It was basically a mechanical device placed on the taffrail which depended upon its action on a vane, fly or helical rotor in the water. A similar idea had been proposed in 1688 by Robert Hooke to the Royal Society but had not been developed (fig. 99).

The first reliable patent log was made by Edward Massey and was introduced in 1802. It was in use until about 1846 when a recording device located on the taffrail, completely divorced from a water-immersed rotor, was invented, the basis of that which is in use at the present time.

Fig 100
Dutch log and perpetual calendar, not signed, *c.* 1759, 153 × 45 × 32 mm (6 × 1¾ × 1¼ in.). Brass tobacco-box with rounded ends.

On the top is engraved a perpetual almanac and lunar calendar with the date 1759. On either side of this, at each end of the box, are portraits representing Julius Caesar and Pope Gregory XIII with the dates 45 B.C. and 1582, these being the dates of introduction of the Julian and Gregorian calendars. On the bottom is engraved a head representing Amerigo Vespucci (?) and the date 1497. Beneath this is a table for calculating the speed of a ship against the time taken by a chip of wood tossed over the side to pass the distance between two marks on the vessel's side. Tobacco boxes of this kind were devised by Pieter Holm, a Swedish mariner, who set up a navigation school in Antwerp in the middle decades of the eighteenth century. The slogan on the side of the box, 'Regt Door Zee', refers to this school.

A useful addition to this equipment was the Dutchman's log box, a brass and/or copper tobacco box of traditional rectangular shape with rounded ends, with a log scale and perpetual calendar engraved upon it. Invented by Pieter Holm, a retired seaman in about 1750, these were made for navigators. Each box is dated at the end of the perpetual calendar (fig. 100).

I have had tobacco boxes made in Britain and Sweden with similar information engraved upon them. The Swedish box bore symbols for the festivals on the perpetual calendar, a plough for sowing, a sickle for harvest, and for Christmas there were three consecutive drinking horns.

The Traverse board

The traverse board was an instrument for dead reckoning used mechanically, i.e. without arithmetic, and was in use from the sixteenth century up to the nineteenth century, and even into the twentieth century. Made of boxwood, the circular face was marked out like a compass with thirty-two points, and in each point there were eight holes to record the course during each half hour of the watch, by inserting a peg into the appropriate hole according to the direction followed at the helm. If, for instance, the course for the first half hour was S.E. a peg was placed in the appropriate bearing in the hole nearest the centre. Say the course was changed to S.E. by E. during the next half hour, then the next peg would be placed in the right bearing in the second hole from the centre, and so on until the end of the watch. Below the face, the rectangular section bore another series of holes in which the speed of the vessel was recorded. Usually reckoned in leagues, the left-hand holes recorded the units and the right-hand holes either quarters or tenths depending on the type of board (fig. 101).

Calculating longitude through astronomy

Although the Portuguese had been able to establish a latitude fix within thirty miles since the second half of the fifteenth century, an effective method of determining longitude was not confirmed until the eighteenth century.

In 1514 John Werner proposed using lunar distances, an argument repeated by Peter Apian in 1524, Gemma Frisius in 1530 and Pedro Nuñez in 1560. When Galileo discovered Jupiter's four largest satellites in 1610, he suggested that they could be used for determining latitude. Although he proposed this information to Philip III of Spain in 1616/17 and to the States General in 1636, his calculations were incomplete at the time of his death in 1642.

Louis XIV of France founded L'Academie Royale de France in 1666 and the Paris Observatory the following year to find a system for establishing longitude. Observations by Picard and Cassini led to the publication of *Connoissance des Temps* by Picard in 1679 which was the first edition of ephemerides to include the eclipses of Jupiter's moons.

The British Royal Observatory was founded 'within our park at Greenwich upon the highest ground' (Royal warrant) in 1675 by Charles II with the prime intention of improving astronomical scolarship to aid the mariner. John Flamsteed (1646–1719) was appointed the first Astronomer Royal, 'With the utmost care and diligence to the rectifying of the tables of the heavens, and the places of the fixed stars, in order to find out the so much desired longitude at sea, for the perfection of the art of navigation.' In 1687 Newton's theory of universal gravitation made possible a precise theory which could make lunar observations useful for longitude.

Because the moon's motion around the earth causes its position to change rapidly in relation to the fixed stars and planets, it could be used as a timekeeper. If there were a method of forecasting the position of the moon at the prime meridian, then the observer at another geographical position would be able to calculate the difference in time and his longitude accordingly.

The exact position of the moon in the heavens can be calculated at certain times: (1) occultations of stars and planets by the moon (i.e. their being obscured by the moon), (2) beginning and end of eclipses (sun and moon), (3) calculation of distance between the moon and a chosen star, called a 'lunar distance'.

Eclipses had been used since the time of Hipparchus (c. 150 B.C.) to determine longitude, when two observers noted the local time of the occurrence and later compared notes on the difference. Explorers followed this procedure and hoped that they would find an astronomer at home who had observed the phenomenon and noted the time, which could then be compared with the local time and the longitude of new-found territory established.

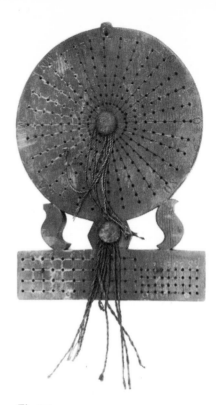

Fig 101
North European traverse board, not signed, early nineteenth century (?), carved and painted wood. Circular board with 32-point compass-rose each point with eight holes and a further eight between the points. A peg could be placed in a hole every half hour in order to track the direction in which a ship has travelled. At the lower end of the board are four further series of thirteen holes each used for reckoning the speed of the ship.

Fig 102 *top right*
English repeating and reflecting circle, signed 'W. & T. Gilbert LONDON'. c. 1820, diameter 271 mm (10¾ in.), brass, with ebony handle and platinum scale.

Flat ring with inset double degree scale (0–360–0) reading to 28' of arc. Carrying this ring is a squared cross strut frame rigidly attached to a central hub through which passes an axle, carrying at one end the object mirror and at the other the three index arms which rotate over the degree scale. All three indexes have verniers with reading microscopes, only one being fitted with tangent and slow motion screws. Rigidly attached to the hub on the back of the instrument are three arms carrying respectively the sighting telescope, horizon mirror with four shades, and the handle.

Fig 103
English quintant, signed 'Cary London', c. 1830, radius 250 mm (9⅜ in.), brass, with platinum scale.

Three arm frame with index. The scale is divided 0°–300° read against a vernier with tangent and clamping screws; one set of three coloured filters for the index glass.

The occultation of the stars method was tried by Edmond Halley during his two voyages in pursuit of variation in 1698–1700, but although only a good telescope and a quadrant were required for the observations for local time, the lunar tables he used by Thomas Street published in *Astronomia Carolina* in 1661 were insufficiently accurate forecasts of prime meridian time. This method, even with accurate tables, was not practical for navigators as the telescope used had to be powerful and therefore large, which would make it uncontrollable in shipboard conditions.

The best method of the three was the computation of the distance of the moon from a selected fixed star. This method was first suggested by Johann Werner in 1514, who proposed to measure the distance with a cross staff, but he was before his time, for the cross staff was insufficiently accurate for the operation and no lunar forecasts existed.

The Repeating and reflecting circle

By the eighteenth century the situation had changed. The Hadley quadrant, an excellent instrument for taking accurate angular distances up to 90°, was familiar to astronomers and seafarers, but while it could be used with a back sight to give a reading up to 180° required for lunar distances, it was relatively cumbersome, awkward and inaccurate.

When Professor Tobias Mayer of Göttingen produced his lunar tables in 1752, he proposed to make an instrument based on the same principles as the octant, but in such a manner as both mirrors could move. By moving them alternately it would be possible to take a number of altitudes successively, so that the instrument added them together, and finally the angle read off the scale could be divided by the number of observations to obtain an average. The error of reading-off could only occur once and would be divided by the number of observations and therefore minimized.

Other versions of the instrument were developed some using the repeating principle and others with three verniers for extra accuracy enabling reading of three angles on arcs of the circle 120° apart. Edward Troughton (see p. 81) was particularly active in this field. The Troughton circle (fig. 102) was an excellent instrument embodying all the refinements the observer required, and, while intended for lunar distances, it was so accurate in computing altitude it was preferred to the octant by some navigators.

Professor Mayer sent a copy of his lunar forecast tables to Greenwich in 1755 and these were checked at the Observatory. They proved capable of forecasting the position of the moon within one minute of arc. It was with some excitement then that the fifth Astronomer Royal, Nevil Maskelyne, who had previously complained of the omission of lunar observations in the Flamsteed records, embarked for St Helena to observe the transit of Venus (across the sun) in 1761 with Mayer's lunar tables. The success of the mission caused the publication of his book which appeared in 1763, entitled *The British Mariner's Guide*, explaining how longitude could be discovered with lunars. The first Admiralty *Nautical Almanac* (fig. 104) was published in 1766 for the year 1767, and amongst the usual Ephemerides and other astronomical information, there were lunar distance tables showing the measurement between the moon and seven bright fixed stars. Apart from the last item, the same astronomical information had been published by the French government in *Connoissance des Temps* since 1679.

In practice the taking of lunar distances was fraught with difficulties. It was at first considered that the observer should have three assistants, one for taking the altitude of the moon, the second that of the star and the third noting the time. He himself computed the lunar distance. This procedure was repeated five times and an average reading was recorded.

It seems a complicated formula, when the apparent procedure would be to compare the local observation with that calculated from Greenwich. The reason is that the distance observed is not the *actual* distance because of two major phenomena, refraction and parallax. Atmospheric refraction

Fig 104
The *Nautical Almanac and astronomical ephemeris, for the year 1769*, published by order of the Commissioners of Longitude, London, 1768. 8vo, 5 pp plus 165 pp. with the calendar page for each month printed in red and black.

Compiled by the Astronomer Royal, Nevil Maskelyne (1732–1811), the *Nautical Almanac* was primarily intended for use in the determination of longitude by method of lunar distances. It remained essentially unchanged from the form given it by Maskelyne until 1834. The third edition for 1769 is of particular interest since a transit of Venus occurred on 3 June. Maskelyne's *Instructions . . .* for observing this phenomenon is bound, separately signed and paginated, at the end of the volume and includes an account of the use of a twelve-inch astronomical quadrant by John Bird, for taking altitudes.

Fig 105
English marine chronometer, signed 'WILLM. FARQUHAR, KING ST. TOWER HILL, London' number T/8, brass and steel.

Eight-day chronometer with silvered dial up and down indicator and subsidiary seconds. Earnshaw spring detent escapement, with Earnshaw type compensated balance. Gimballed in brass-bound box.

causes the bending of the line of sight as it passes through the earth's atmosphere, so the true position of a celestial body is always lower than it appears to be.

When the altitude of a celestial body is taken, it is calculated as if from the centre of the earth, but the observer is standing *on* the deck of a ship slightly above the surface of the earth; creating an error known as parallax. The positions of the star and the moon must therefore be corrected for their true positions, a ploy which is called 'clearing the the distance'.

Basic instructions for corrections to be applied for all the aberrations were given in nautical almanacs, as they still are to this day, but the complicated calculations were beyond the uneducated seaman. Lunar distances were quoted in the British *Nautical Almanac* until the year 1906, although apart from the knowledge that they could be used to check chronometrically determined longitude values, the method was little used.

Chronometers

The problem of calculating longitude precisely was finally solved when John Harrison (1693–1776) invented a chronometer sufficiently accurate to comply with the limits set down by the Admiralty, for which, in spite of considerable opposition, he won the £20,000 award. This Yorkshire carpenter-turned-clockmaker's brilliant conception of a clock which could keep accurate time under sea-going conditions has already been more than adequately described in print by R. T. Gould in *The Marine Chronometer, its History and Development* (1923) – and further reading is suggested in the bibliography.

The invention became indispensable, and an extract from a famous log-book, that of Cook's second voyage, reads:

'Longitude from Mr Wales's Observations.
By the moon and star Aquilae   5° 51′ ⎫   Mean 6° 13′ 0″
By the ditto and do. Aldebaran 6   35 ⎭
By Mr Kendal's Watch   —   — 6·53$\frac{7}{8}$

Larcum Kendal (1721–1795) was the maker of K.1, the watch referred to in the log which was checked by Cook for the Admiralty on his second and third voyages (1772–5 and 1776–9), and also by George Vancouver in his survey of the west coast of North America (1791–5). Made as a duplicate of Harrison's No. 4 it proved as admirable as its master. Kendal made two more chronometers with amendments of his own, but these were inferior. K.2. accompanied Bligh on the *Bounty* on the Breadfruit expedition in 1787 at the time of the mutiny, but the use of it was denied him when he was cast adrift in the longboat, as it was said that with it he would be able to get home. Acquired in 1808 by an American whaling captain who called at Pitcairn, it was stolen from him in South America. It turned up again in 1840 when it was bought for fifty guineas by Captain Thomas Herbert R.N. who brought it home in 1843.

The Admiralty lent it to the Royal United Services Institution and when a few years later they proposed to sell it, they discovered it was not theirs to sell. It was returned to the Admiralty, who placed it at Greenwich where it now takes its place in the National Maritime Museum alongside K.I. and H.I., H.2., H.3., and H.4.

Amendments to the chronometer were made by John Arnold and Thomas Earnshaw, and these remained the basis of the instrument throughout the nineteenth century (fig. 105). In France chronometers made by Pierre le Roy and Ferdinand Berthoud were given sea trials in 1766, 1768 and 1771, and the latter so improved his model, that he was able to make and sell a large number. The use of chronometers became so widespread that they were actually issued by the Admiralty to ships of the Royal Navy, from about 1825. In general use by the 1880s, they superseded the use of lunars but it was not until after 1906 that lunar tables were omitted from the *Nautical Almanac*.

Fig 107
English dipleidoscope, signed on the cap 'Dent's Patent Meridian Instrument 61, Strand, London.' and on the side 'E. J. DENT PATENTEE 800', *c.* **1870, 71 × 49 mm (2¾ × 2 in.).**

Dent's early form of dipleidoscope. The instrument consists of a hollow triangular prism, in which an image of the sun is reflected from two of the plates onto the third. Since the dipleidoscope is fixed, these two images will coincide at solar noon. The instrument can also be used to determine the meridian transit of other celestial bodies.

Fig 108
English (?) navigational calculator, not signed, *c.* **1800, diameter 317 mm (12¼ in.), brass, in original mahogany box.**

A complex device, possibly a prototype, for solving problems of spherical trigonometry relating to great circle sailing whether of distance, direction, latitude, longitude or hour angle to an accuracy of 15′ of arc. The circles of which it consists are, reading from the centre outwards: equator marked with hour angle I–XII/I–XII = longitude; meridian ring for celestial body 90°–0°–90°; observer's meridian with latitude scales 85°–0°–85°; zenith distance great circle 0°–180°–0°.

The compass discs are used to represent the declination and hour angle of the celestial bodies under consideration, and the position of the observer's zenith. The whole instrument may be set up for use in its box. The bearings and altitudes of celestial bodies may be worked out using the azimuth ring and the half circle.

Meridians

The last thing to determine was a common prime meridian. According to W. E. May in *A History of Marine Navigation*, the first chart to use Greenwich as the prime meridian was in 1738 by Fearon and Eyes, but when the first *Nautical Almanac for Lunar Distances* was published in 1767 using Greenwich as the prime meridian and lunar distances became increasingly used for longitude, Greenwich became the standard for England and America.

Other nations were using their capitals as their prime meridian and were loth to relinquish the custom, but at a geodetic conference in Rome in 1883 delegates were recommended to accept Greenwich, but refused to do so, and it was not until the following year at another conference convened by the U.S. Government in Washington that the Greenwich meridian was adopted by twenty-one nations to one (France and Brazil abstaining) as the prime meridian of the world. The meridian used is that which passes through the cross-wires of the telescope in the Airy Transit Circle in the Royal Observatory, and is indicated outside the building by a brass strip across the courtyard.

The reasons for the collector's fascination with navigation instruments is deeply psychological. These objects were no toys for gentlemen but were working tools which saw active service in dangerous conditions, and were cherished by their owners as a means of survival. The strong feelings engendered by these instruments also lie in the romance of the sea, now that time has softened the appalling conditions to which seafaring men were exposed. The rugged individualism of the sailor scientists appeals to men (and women) trapped indoors in the machine of modern business, and in handling their artifacts a fresh breeze seems to blow open the door to freedom.

Fig 109
Hadley quadrant by Goater, mid-eighteenth century.
See also Fig. 88.

Fig 110
English backstaff by John Goater, Wapping, c. 1750.

Fig 111
British mariner's equinoctial dial,
not signed, 1634, radius of dial
250 mm (9⅞ in.), slate.

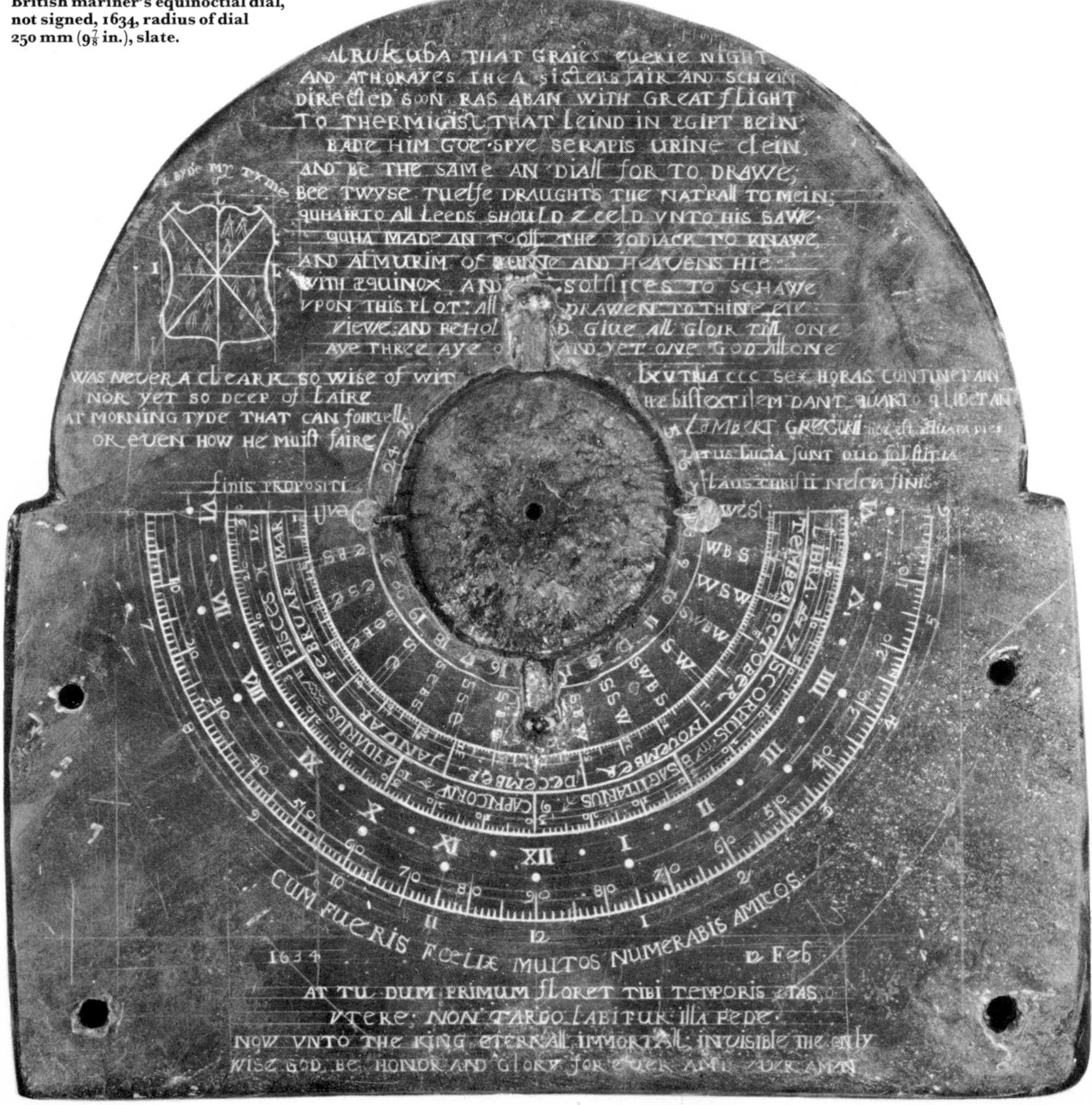

Reverse of fig 111.

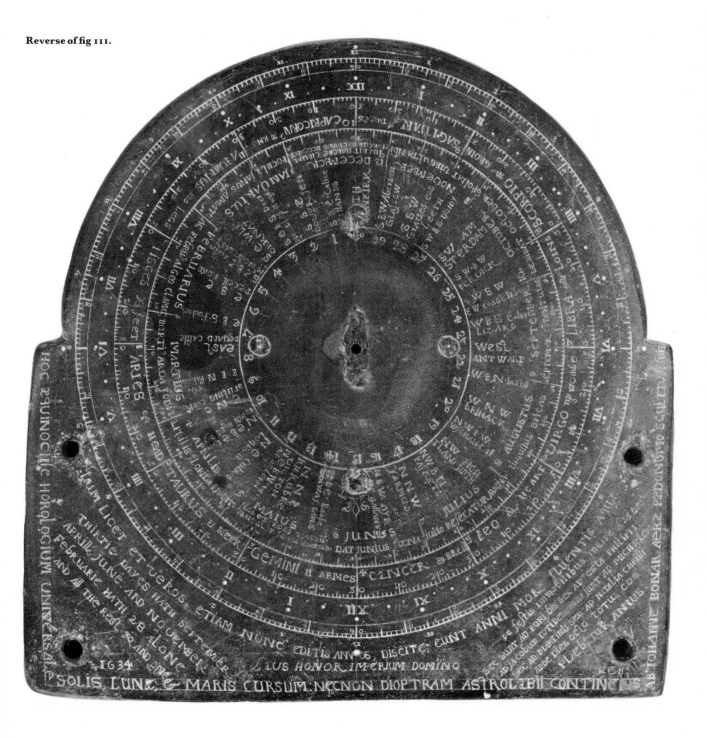

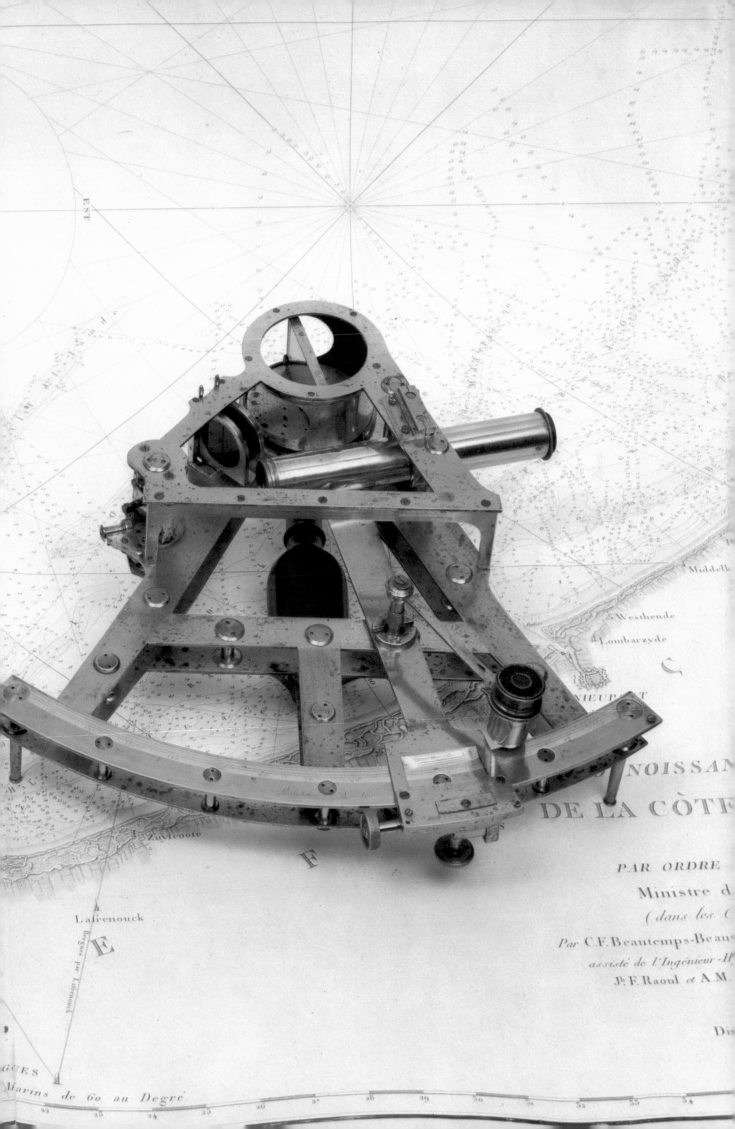

Sundials

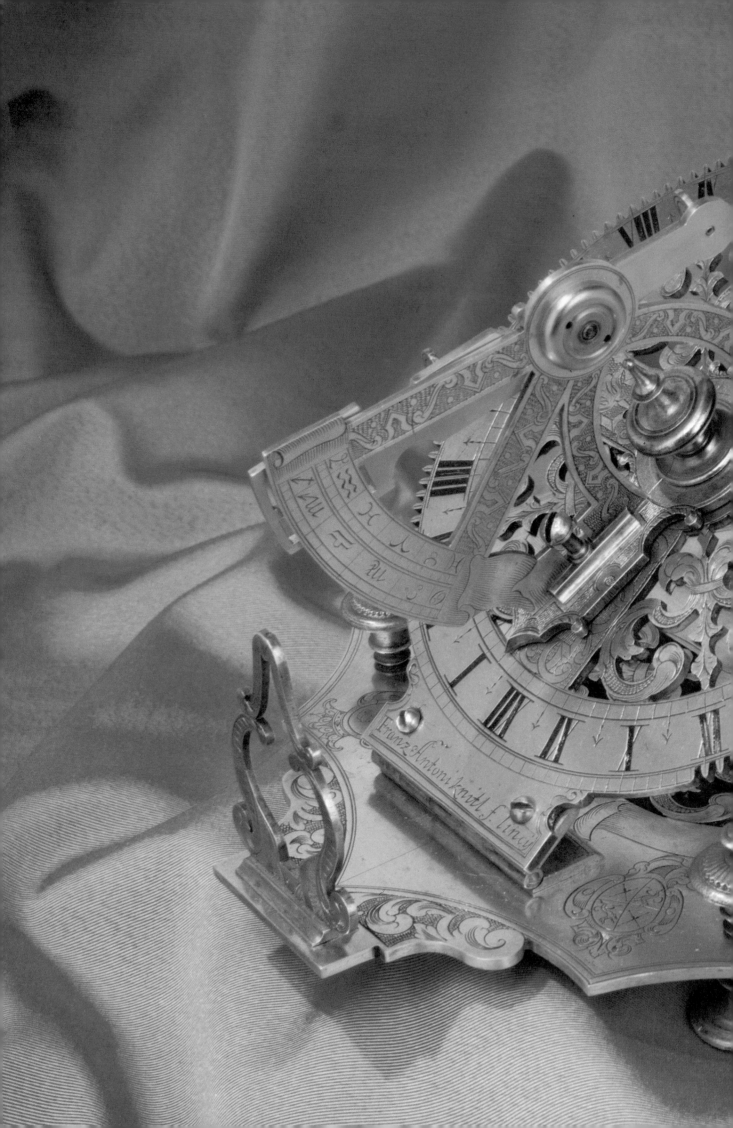

N ancient instrument which may be made in a great number of
pleasing and beautiful forms, the sundial offers variety,
geometrical ingenuity and fine craftsmanship. Although the
history of the sundial begins in the ancient world of Egypt,
Greece and Rome, the great age of dialling (as the art of
making sundials was known in England) fell in the sixteenth, seventeenth
and eighteenth centuries, and it is chiefly the instruments produced in
that period that are described here.

These are the instruments which have survived in greatest numbers.
Medieval dials are extremely rare, but the profusion of styles and the
ingenious variations concocted by the instrument-makers of the seventeenth and eighteenth centuries more than compensate for their comparatively recent date. The origins of these dials, often appearing
apparently independently in different parts of Europe, are very obscure
and do not follow any obvious evolutionary line. It is frequently impossible
to be precise as to their place or date of origin. What is more certain is
that they become far more widespread in the seventeenth and eighteenth
centuries, for this period saw a rise in popular understanding of mathematics, and a great surge of interest in technology and popular science,
which coincided with the development of relatively accurate clocks and
watches. The combination of the two led to a much wider interest in time,
the methods by which it is reckoned and kept, and also in problems of
accuracy. For checking the performance of a clock, however, there was
no standard except that of the sun. A sundial was therefore essential for
the purpose. Far from reducing the usefulness of the sundial, the mechanical clock probably stimulated its production and use. Moreover, the
notorious unreliability of clocks in a period when the growth of commercial activity made punctuality steadily more important helped to
stimulate demand for small, portable and easily used sundials. The
instrument-makers quickly produced dozens of types to meet demand.
Larger sundials, designed to be fixed in a garden, on the wall of a building,
or kept in one location were also produced in quantity.

Whether it be fixed or portable, designed for use in one latitude or in
several, the basis of a sundial is that time is measured by the position of the
shadow, or of a spot of light, thrown by an indicator (the gnomon) upon

Fig 113 *(on previous page)*
Mechanical equinoctial dial by Franz
Antonij Knitl, *c.* 1700. See also
Fig. 140.

Figs 114 & 115
German perpetual calendar and
aide-memoire, signed 'Joh: Georg
Mettel Fecit', mid-eighteenth
century, 100 × 52 mm (4 × 2 in.),
silver with ivory leaves.

Two covers jointed at one end only;
the upper cover carries a volvelle
with scales for the length of the
month, feast days and festivals, date
and length of longest day in the
month, position of the sun in the
zodiac, length of day and night and the
times of sunrise and sunset.
Surrounding the volvelle is flowing
foliage decoration encircling
cartouches with images representing
spring and summer and with the
relevant zodiacal symbols. On the
back cover is a second volvelle for the
age and phase of the moon, with in the
centre a perpetual calendar.
Surrounding the volvelle is flowing
foliage decoration similar to that on
the top cover and encircling
cartouches for autumn and winter.
The top edge of the instrument is
scalloped and incorporates a shell
motif. With four leaves.

106

a scale of hours. The time is indicated either by the length of the shadow cast, which depends upon the sun's height above the horizon (altitude dials), or by the direction of the shadow (direction dials). This depends on the position of the sun measured either in the equinoctial (hour-arc) or along the horizon (azimuth). There is also a number of dials in which the position of a separate indicator against the hour scale shows the time. Altitude dials are generally more suitable for use in lower latitudes where variation in the sun's height throughout the day is more marked than in northerly regions. All altitude dials, however, require compensating for the change in the sun's declination throughout the year and include a scale for this purpose.

Fig 116
German ring dial, signed 'IOANNES : GEORGIVS : ZECH : ANNO 1699', diameter 45 mm (1¾ in.), brass.

Brass ring with suspension mount and chain. The inner surface is engraved with an hour scale and repeats the maker's initials and the date. The outer surface is a calendar scale. The central portion of the ring may be slid round to adjust the sighting hole for declination against the calendar scale.

Fig 117
English ring dial, signed 'L. PROCTOR SHEFFIELD', mid-eighteenth century, diameter 110 mm (4¼ in.), brass.

The interior is engraved with three sets of hour lines, a declination scale is engraved on the exterior against which the central sleeve may be set; with brass suspension loop.

Altitude dials

Ring

Probably the commonest, and certainly the simplest, form of altitude dial was the ring dial (figs. 116 and 117), in which a ray of sunlight passes through a small hole in the side of a ring set vertically and indicates the time on a scale of hours engraved on the interior of the ring. In this form the dial will be correct for only one value of the sun's declination. Various expedients were adopted to overcome this shortcoming such as engraving separate scales on the two edges of the ring, one for the winter months, the other for the summer, or by having two sighting holes, one on each side of the suspension-piece. Most commonly, however, the hole was placed in a sliding collar at the centre of the ring which could be adjusted for declination against a calendar scale. The accuracy of these dials was generally poor.

Pillar

Pillar or cylinder dials (fig. 120), commonly made of wood, more sumptuously of ivory and very rarely of brass, consist of a short column on which hour lines are engraved. Round the top of this column (occasionally elsewhere) is a calendar scale against which the gnomon is set. The gnomon can usually be folded back into the body of the instrument when not in use. The whole dial is suspended with the gnomon facing towards the sun and the tip of the shadow indicates the time. Dials of this kind continued in use among peasant communities at least until the beginning of the twentieth century, and are sometimes referred to as 'Shepherd dials'.

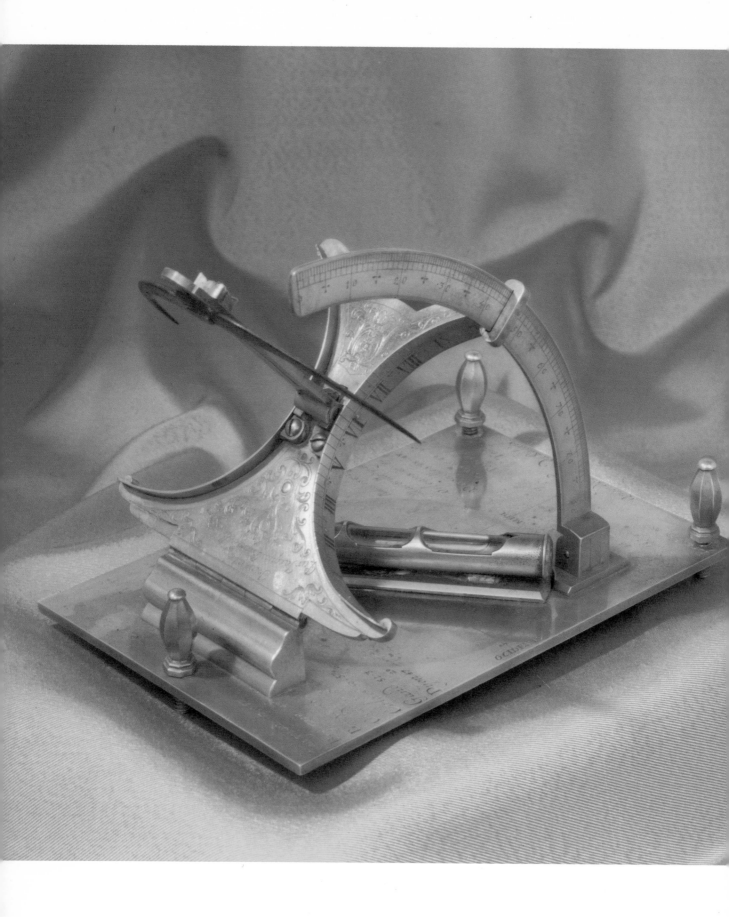

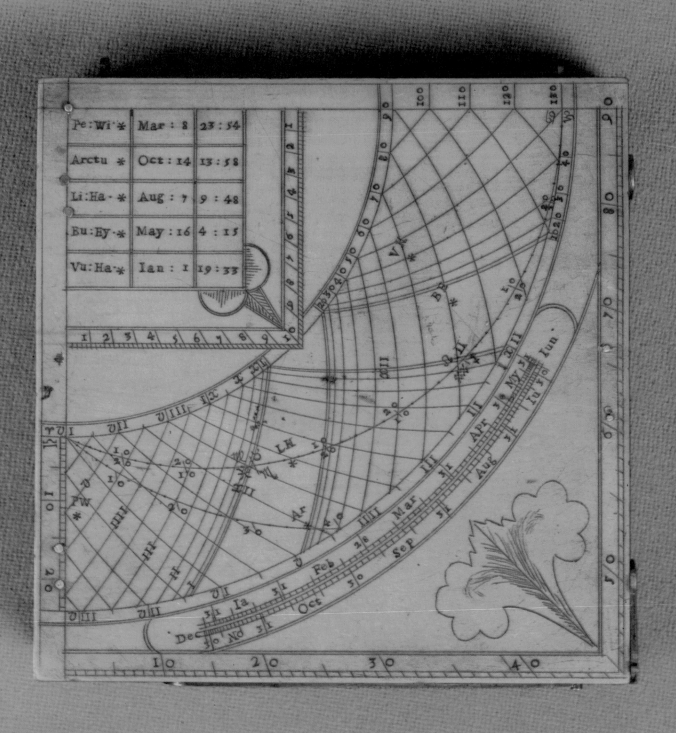

Vertical plate

A dial which works in the same way as the pillar, but is considerably rarer, may be drawn on a flat plate, the gnomon being made adjustable against a calendar scale along the top edge. In this form the instrument is known as a vertical plate dial.

Scaphe

One of the oldest known forms of sundial of which large marble examples survive from the Roman period, the scaphe consists of a hollow sphere on which the hour lines are drawn, in which there is a central gnomon. The time is indicated by the position on the hour scale of the extreme tip of the shadow. In form the scaphe shows great variation ranging from a complete hemisphere to a shallow spherical cavity such as is found in some diptych dials (see below). Dials of this type are also found engraved on the interior of chalices, goblets and spoons.

Quadrants

At least three forms of quadrant which could be used for time-finding were known in the Middle Ages. They were known in the West as the *quadrans vetustissimus* (the oldest quadrant), the *quadrans vetus* (the old quadrant) and the *quadrans novus* (the new quadrant). The latter was devised in the thirteenth century by the Jewish scholar Prophatius Judaeus of Montpellier. It continued in use in Islamic regions almost until the present century, being found on many nineteenth-century Turkish quadrants. In this context it is often found together with a sinecal quadrant which derives from the *quadrans vetustissimus* (fig. 122). The *quadrans vetus* also continued in use throughout the seventeenth and eighteenth centuries on both Islamic and Western instruments, being found frequently as an auxiliary scale on astrolabes and on other types of quadrant (figs. 123). In sixteenth- and seventeenth-century Europe many ingenious variations of the horary quadrant were developed, of which the Gunter quadrant, the most popular of them in England, has been discussed in chapter 1 above.

Rojas or geminus

A form of sundial sometimes found on the back of nocturnals, quadrants or, more rarely, astrolabes is the vertical universal dial known as the Rojas dial. This employs an orthographic projection onto the plane of the meridian of the hour circles and parallels of declination. The projection is similar to that used for one form of universal astrolabe and was first described by the Frisian Hugo Helt in the *Commentarii de astrolabii* of

Fig 120
French pillar dial, not signed, nineteenth century, overall height 95 mm (3¾ in.), softwood.

Fig 121
French or German (?) vertical plate dial, not signed, late seventeenth century (?), brass.

Circular disc to which is attached an adjustable fan-shaped tab with suspension ring engraved with a calendar scale. The tab can be

adjusted against a pointer fixed to the disc. On the lower edge of the face is an hour scale, the divisions being radial from the short folding gnomon on the upper edge. The centre of the dial is occupied by a coat of arms with the motto 'MILITANDUM'. On the back is a table for simple multiplication and addition.

Fig 118 *(previous page, left)*
Crescent dial, signed Baradelle, c. 1800. See also Fig. 149.

Fig 119 *(previous page, right)*
Gunther quadrant ascribed to Henry Sutton, c. 1650.

Fig 122
Turkish quadrant, not signed, eighteenth century, radius 160 mm (6¼ in.), boxwood, the two faces of the instrument are lacquered yellow and the edges red; the lines, script and numerals are marked in black, red and gold.

Brass inset at the apex for double plumb line (now missing). On the front is a Prophatius astrolabe quadrant, on the back a sinecal quadrant with arcs of sines, versed sines and the arc of the obliquity of the ecliptic. A free translation of the inscription reads 'May Allah bless my father, his children and all the faithful.'

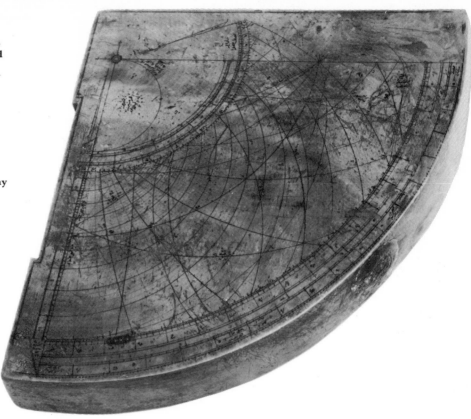

Fig 123
Italian (?) quadrant, not signed, radius 405 mm (16 in.), brass.

On the face is a *quadrans vetus* with sliding calendar scale for declination and a shadow square; on the reverse is a Rojas projection of the sphere similar to that in figure 21. The quadrant was originally equipped with a plumb line and sliding bead.

A finely made modern copy.

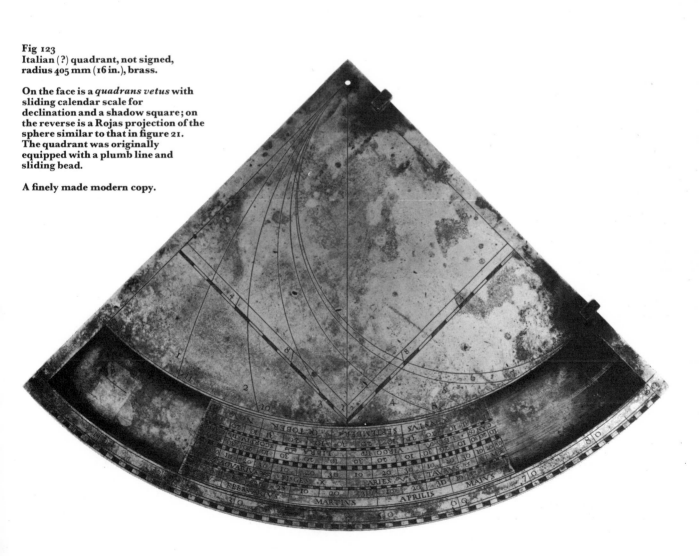

Ferdinand de Rojas (1550), although examples of the sundial projection are known which pre-date this work.

The projection is usually engraved on a disc which revolves about its centre, and which is set for latitude against a scale engraved on the plate behind. Once set for latitude the plate is held vertically with the sights directed towards the sun, and the intersection of the plumb line with the appropriate parallel of declination gives the time. When engraved upon a quadrant (fig. 19) or astrolabe the sights, and (in the case of the quadrant) the plumb line fitted to the instrument itself, may be used. For use on a nocturnal, however, a triangular arm has to be added at the centre to carry the sights and a plumb.

Direction dials

Horizontal plate

The horizontal plate dial is the most common, and one of the most attractive, of all dials. The hour scale is engraved on an horizontal plate, and the lines radiate out from a point from which the gnomon rises. The shadow-casting edge of the gnomon is parallel to the earth's axis, and, if the dial is correctly oriented, the position of the shadow against the hour lines shows the time. Large dials of this kind, made of heavy brass, bronze or slate for outdoor use in one fixed position, are quite common. In the eighteenth century some of the larger examples were engraved along the circumference with a scale for the equation of time. Small portable dials of this kind, intended for use only in one latitude, are also found and these usually incorporate a small compass for orientation (figs. 125, 126). It is one of the paradoxes of history that the small circular wooden dials of this kind were among the cheapest and most popularly used of dials but are now among the rarest of all.

String-gnomon

A portable form of horizontal plate dial was apparently very popular in the sixteenth and early seventeenth centuries. In it a hinged lid or other vertical support is added to carry a string which takes the place of the gnomon. The lower end of the string is attached to the plate. If the upper end is attached to a slide, the dial can be adjusted for use in various latitudes. In a further form, three or four holes are drilled in the lid as a crude form of latitude adjustment (figs. 128 and 129).

Butterfield

An adaptation of the horizontal dial which is supposed to have been introduced by Michael Butterfield, an Englishman who worked in Paris from *c.* 1685–1724, the Butterfield dial is both portable and universal. A small octagonal or oval plate has engraved on it several concentric hour scales for different latitudes, and has a small compass inset. A folding gnomon is made adjustable for latitude against a degree scale engraved on its side. The support for the gnomon is made in the form of a bird, the beak of which acts as the indicator upon the degree scale. Such dials, which were sold in close-fitting fish-skin boxes, were extremely popular throughout the eighteenth century (figs. 130 and 131).

Compass

This was a popular dial, probably of medieval origin. A cut-away dial plate with a folding gnomon is placed above a compass mounted in a brass, usually circular, box. A screw-on lid completes the fittings. Dials of this kind were seldom made adjustable for latitude (figs. 135 and 132).

Magnetic compass

A small dial, chiefly made during the eighteenth and nineteenth centuries, in which the floating card of a magnetic compass is marked with hour lines, and also carries a gnomon. In principle the dial is self-orienting as the card will align itself on the meridian and so does not need to be pre-set.

Fig 124
Compendium by Hans Ducher, 1593.
See also Fig. 157.

Fig 125
English horizontal plate dial, signed
'Made by Tho Wright Instrument
maker to his MAJESTY', *c.* 1730–40,
diameter 368 mm (14½ in.), brass. Hour
scale with *fleur de lys* decoration and
coat of arms (not identified) of the
original owner; equation-of-time
scale and 16-point compass-rose;
shaped and pierced gnomon.

Fig 126
English horizontal plate sundials,
signed:

Left
  'Cole Maker in Fleet St. London',
diameter 298 mm (11¾ in.), *c.* 1760

Centre
'W. & S. JONES Holborn London' 203 mm
(8 in.), *c.* 1820.

Right
'Heath & Wing London', diameter
298 mm (11¾ in.), *c.* 1760.

Fig 127
German horizontal dial, not signed,
c. 1600, 56 × 68 mm (2¼ × 2¾ in.).
Single base plate engraved on the back
with the latitudes of forty-four towns
and with inset compass *(below right)*.

Overlaid on the base of a circular
silver hour scale engraved in the
centre with a representation of Time
pursued by Death. Around the rim of
the compass box is the slogan 'Hora
Notissima Tempora Pessima
Vigilamus' (We watch for the hour
time makes the worst). A spring-
loaded style pivots from a position
immediately above the compass
being set against a latitude arc
(0°–70°) hinging from the centre of
the hour scale.

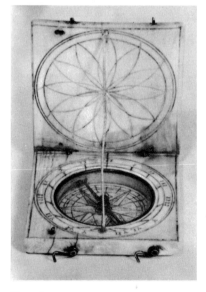

Fig 128
French or German (?) string-gnomon
sundial, not signed, sixteenth century,
ivory with brass catches and hinges.

Inset compass with 8-point rose,
surrounding the compass aperture is
the hour scale. Inscribed on the lid is a
formal geometrical flower design
with twelve petals.

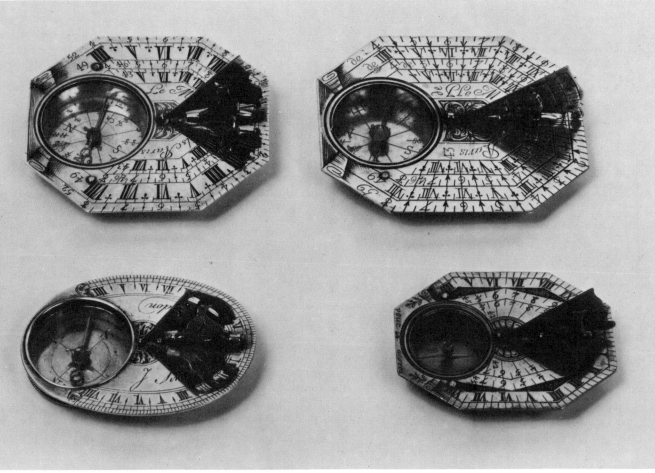

Fig 129 *far left*
French string-gnomon dial, not signed, *c.* 1600, lignum vitae and ivory with silver clasps, 74 × 60 mm (2⅞ × 2⅜ in.).

Oval box containing a glazed compass with the cardinal points painted in gold and black on a blue ground, surrounded by three hour scales for latitudes 42°, 45°, 48°. The inside of the lid is decorated with formal leaf and scroll designs and has a table of the latitudes of thirteen places together with spring and slide attachment for the string gnomon, adjustable against the latitude scale.

Fig 130 *left*
French Butterfield dial, signed 'Butterfield A Paris', *c.* 1700, 58 × 50 mm (2¼ × 2 in.), silver with original velvet-lined fish-skin case.

Octagonal base, engraved on the under side with the latitudes of various places; spring-loaded folding gnomon engraved with a degree scale for latitude 40°–60°; four hour scales for latitudes 43°, 46°, 49° and 52°; inset compass.

Fig 131
Butterfield dials
Top left
Signed 'Le Maire Fils A Paris', *c.* 1740, silver, length 64 mm (2½ in.)

Top right
Signed 'P. le Maire A Paris', *c.* 1750–60 (?), silver, length 69 mm (2¾ in.) Both these dials were probably made by Pierre Le Maire.

Bottom left
Signed 'J. Simons London', late seventeenth/early eighteenth century, silver, diameter of major axis 59 mm (2¼ in.). An unusual example made for use exclusively in England. On the base are the names and latitudes of six English towns; the degree scale on the gnomon (engraved on one side only) is 50°–70°; single hour scale.

Bottom right
Signed 'Sevin à Paris' *c.* 1680 (?), silver.

In practice such dials are extremely unreliable since it is improbable that the magnetic axis and the meridian will coincide. Moreover, no correction was made for changes in magnetic deviation.

Inclining

From at least the beginning of the eighteenth century, English instrument-makers were producing a horizontal dial similar in appearance to a compass dial, but with the dial-plate hinged at the north point, and adjustable against a quadrantal scale of degrees of latitude. Such dials work in the same way as a normal horizontal dial except that they may be used in more than one latitude. If the dial was removed from the latitude for which it was made, the dial plate could be adjusted against the degree scale by a sufficient number of degrees for it to remain parallel to the horizontal plane of the original place. Dials of this kind are often elaborate and heavily made. They may incorporate spirit levels and adjustable feet for the compass plate, with clamping screws to hold the dial in position. Although portable, they were clearly intended to sit handsomely in a window as a fine adornment to an elegant room (figs. 133 and 134).

Vertical

Vertical dials are nearly always fixed, and were often drawn directly onto a wall, but on occasion they are found made of brass or slate. The appearance of the dial varies according to the direction (aspect) faced by the wall which it is intended to adorn (fig. 136). Such dials are also, but more rarely, found drawn on glass for incorporating into a window, when the dial is drawn in reverse since it is intended to be read from the inside.

Polar

This dial is usually found on polyhedral and diptych dials (see below) in which the plane of the hour scale is parallel to the earth's axis. As a single dial it is of extreme rarity.

Spherical equinoctial

All equinoctial dials have their hour ring in the plane of the equinoctial circle. Since the shadow of the gnomon moves round this at a uniform rate the ring may be divided with great accuracy. In its spherical version the hour circles are drawn on a solid sphere often made of marble. A gnomon is not required, since half of the sphere will be permanently in shadow and once the sphere is properly set with its axis parallel to the earth's axis, the edge of the shadow will indicate the time.

Equinoctial

Often attractively made in gilt and silvered brass this pocket dial was produced in great numbers, by German makers particularly those centred on Augsburg. It was, however, manufactured throughout Europe and shows some variation of form according to its place of origin. It consists of a compass set into a base plate, usually on feet, with an hour ring hinged to the north point of the compass. Pivoted across this ring is a horizontal bar which carries the gnomon. This often runs both above and below the ring so that its shadow always falls upon the scale and can be seen easily. The equinoctial ring is set for latitude against a quadrantal scale engraved on a hinged arm at the side. All the parts of the dial may be folded flat so that it can fit into a thin case (fig. 139).

Self-setting-equinoctial

A variant form of the equinoctial dial probably originating in the late seventeenth century has a dial, pointer, cam, and spring instead of the quadrantal scale for setting to latitude. If the pointer is rotated to the desired latitude the hour ring, when released, will spring up to the correct inclination (fig. 138).

Fig 132
Persian *qibla* indicator, signed 'Made by 'Abd al-[A'imma]', early eighteenth century, diameter 66 mm (2½ in.), brass.

Circular box with hinged lid, containing a compass of which the needle is missing. Above this is a cut-away plate engraved with a semi-circular scale of degrees. Tables for the *inhiraff* and *jiha* of various towns are engraved on the outer and inner surfaces of the lid, and on the outer surface of the base.

An adaptation of the compass dial, this instrument enables the azimuth of the *qibla* (direction of Mecca) to be determined and so the time and direction for prayer.

Fig 133 *below*
English inclining dial, signed 'B Scott Fecit', *c.* 1725, length of side of square 73 mm (2⅞ in.), brass and silvered brass.

Base with inset compass having an 8-point rose with the directions initialled. The edges of the base are decorated with an oak leaf border. Hinged equinoctial plate with folding spring-loaded gnomon, both engraved with flowing leaf decoration. The latitude arc is a replacement.

Fig 134
English inclining dial, signed 'I.
Sisson LONDON', c. 1740, base 212 mm
(8⅜ in.) square, silver.

Plain base with inset compass;
folding latitude strut with double
(reversed) degree scale (0°–90°). The
inclining plate is engraved with a
flowing leaf pattern and carries a
degree scale radial from the base of
the gnomon.

Fig 135 *below*
English compass dial, not signed,
c. 1730, diameter 85 mm (3⅜ in.), brass
with compass papers printed from
engraved plates.

Circular box with compass paper
with 16-point rose, named
directions, and an outer degree scale
in four quadrants 0°–90°–0°–90°
numbered from the north point, in the
lid. In the box is a second compass
paper with needle. Above the compass
is a cutaway brass plate engraved
round the edge with an hour scale and
with a decoratively engraved cross-
strut from which the folding gnomon
rises.

Mechanical equinoctial

This type of equinoctial dial appeared during the late seventeenth and
early eighteenth centuries. A radial pointer with sights is mounted on the
hour ring. The sights are turned towards the sun, and the tip of the pointer
indicates the time on the hour scale. Since this scale is uniformly divided,
a smaller hand may be geared to the hour plate, which is toothed, to
indicate the minutes on a subsidiary dial (fig. 140).

Astronomical ring

The first description of an astronomical ring is in *Annuli astronomici . . . vsvs*
(1558) by Gemma Frisius, who claimed only to have improved the
instrument, not to have invented it. It is essentially a simplified form of
armillary sphere, consisting of four rings. The outer, meridian, ring is
engraved with a latitude scale; hinged to it so that it can open out at right
angles is the equinoctial ring engraved with the hour scale. Pivoted in
this is a third ring engraved with a zodiacal calendar within which slides
a fourth carrying sights. The suspension ring would be set to the appropriate latitude and the sight-carrying ring slid round to the right date. It
would then be suspended and turned so that the sights pointed towards the
sun. The time could be read off the scale on the equinoctial ring. In
addition to indicating the time the instrument could be used for observational measurements of altitude and azimuth. In a later form, popular
with French and German makers in the early eighteenth century, the
sights were often replaced by one or two alidades. The instrument,
especially when made large and heavy, could thus be used as a mariner's
astrolabe.

Equinoctial ring

The invention of this extremely popular and convenient self-orienting dial
is ascribed to William Oughtred (1575–1660) in the early decades of the
seventeenth century. It is basically a simplified form of the astronomical
ring in which the sight-bearing third ring is replaced by a bridge engraved
with a declination scale. A small cursor, which is pierced to supply the
index, slides along the bridge. The user would slide the suspension mount
along the latitude scale engraved on the outer (meridian) ring to the

Fig 136
English vertical dial, not signed, 1740,
440 × 380 mm (17¼ × 15 in.),
sandstone with lead gnomon.

South-facing dial with initials
'I R W' in tulip lettering at the top.

Fig 137
German equinoctial dial, signed
'Gerhard Kloppenburgh Invenit et
Delineavit Gerhard Cremer
Sculpsit', 1714, brass.

Circular plate engraved on each side
with an hour scale, one side being
marked for spring and summer, the
other for autumn and winter. On the
spring/summer side is the motto
'SIC TRANSIT GLORIA MUNDI' inscribed
within a banner. At either end of the
banner are angels swinging censers.
The hour scale is marked with the
names of various places throughout
the world.

Fig 138
German equinoctial dial, signed
'M[ichael] T[obias] Hager a Armstadt',
c. 1690, 45 mm (1¾ in.), gilt brass, silver
and steel.

Circular case in the form of a watch
case with pendant, engraved on the
back with the names and latitudes of
twenty-five places. A compass is
inset in the base and viewed through a
heart-shaped glazed aperture.
Engraved on the base is a circular
scale for latitudes reading from
40–65 with a small index.

Fig 139 *left*
German equinoctial dials, signed by Muller (upper) and Vogler (lower two), brass, *c.* 1700.

Three elaborately decorated examples by well-known Augsburg makers. Each has its original case, in the lid of which a perpetual calendar was sometimes placed.

Fig 140
German mechanical equinoctial dial, signed 'Franz Antoni Knitl, f. Lincii', length 260 mm (10¼ in.), brass and silvered brass.

Octagonal base plate with scalloped shoulders, engraved with nine coats of arms and with four screw feet for levelling. Rigidly attached at one end is a plummet holder. The centre of the base is pierced and decoratively engraved. Along the length of the base is a latitude scale against which the equinoctial arm may be set by a strut and holder. This holder is adjusted by loosening a winged nut below the plate. The hinged equinoctial plate is pierced and engraved in the centre in similar manner to the base and carries a double hour scale. Attached to its centre is a vertical member carrying the sighting arm (a replacement) adjustable for declination against a calendar scale. Also attached to the centre is an arm radial with the equinoctial plate carrying at one end a subsidiary dial with index (a replacement) for minutes which meshes by a pinion below with the toothed outer circumference of the equinoctial plate.

Fig 141
English universal equinoctial dial, signed 'DOLLOND LONDON', late eighteenth century, brass, diameter 63 mm (2½ in.).

Circular brass plate carrying a compass with silvered 8-point rose with the cardinal points marked by initials. Hinged latitude arc engraved 0°–60° against which the hinged and cut-away equinoctial plate may be set. The latitudes of eight places are marked on the underside of the compass box. Original velvet-lined fish-skin box, with brass clasps and hinges.

Fig 142
French universal equinoctial ring
dial, signed 'Butterfield à Paris', late
seventeeth/early eighteenth century.
Diameter 74 mm (2⅞ in.), brass. For
northern and southern latitudes.

The suspension ring passes through a
rotatable eye attached to a mount
which is free to slide over the two
scales of degrees (0°–90°) engraved on
the (outer) meridian ring. This also
carries the names of main towns with
their latitudes. The inside and one
surface of the (inner) equinoctial ring
are divided for hours (1–12 ×2), the
second surface carrying additional
town names and latitudes. The
decorated bridge with cursor, is
engraved with calendar and zodiac
scales for declination. In fitted, satin-
lined, fish-skin case.

Fig 143
English armillary sundial, not signed,
c. 1790, iron with gilt lettering, lead
base.

Outer meridian ring carrying rings
representing the poles, tropics and
equator and two additional rings.
The hour scale is marked on the inside
of the broad equinoctial ring. The
shadow cast upon it by the rod
representing the polar axis will, if the
dial is correctly aligned, show the
time. Surmounting the sphere is a
figure of a lion rampant.

Fig 144
French horizontal dial, inscribed
'A SALINS 1623', 212 mm square (8⅜ in.),
lead.

Square plate with elaborate floral
decoration in the upper corners and
cherubs blowing. In the upper centre
is a circular hour dial with copper
gnomon (a replacement). Below is an
horizontal scale of saint's days and
diagonal lines for the length of the
day. At the lower centre of the plate is
a winged sandglass surrounded by
decorative foliage. In the lower
margin is the inscription in small
capitals, 'LE CONTE DE BOURGOIGNE ET
AUTRES REGIONS D'ENVIRON 47 DEGRES
D'ELEVATION.'

Fig 145
English universal equinoctial ring
dial, not signed, early eighteenth
century, diameter 154 mm (6 in.),
brass.

For northern and southern
latitudes. Engraved on the bridge in
flowing script are the initials 'B H'. On
the back of the meridian ring is a
0°–90° scale for determining the sun's
meridian altitude.

Fig 147 *right*
English equinoctial ring dial, signed
'G. Wright Inv'./ Martin Londini Fecit',
diameter of base 170 mm (6¾ in.),
brass, silvered brass and steel.

Circular base with four levelling
screws as feet and two spirit levels.
Mounted in the centre in a fixed arc
engraved with a vernier scale is a
semi-circular ring supplying the
meridian, engraved with a scale of
degrees (0–90) for latitude adjustment.
Pivoted inside this ring is a second
ring with a cross piece carrying
sights, the point of which passes
through one end of the fixed ring, to
carry an hour scale (1–XII × 2) also
with vernier. In use the inner ring is
turned until the sun shining through a
small hole in the central sights falls
exactly on a line engraved centrally on
the inner surface of the fixed meridian
ring.

Fig 146 *left*
English equinoctial ring-dial, signed
'THO: HEATH LONDON', *c.* 1750, brass and
silvered brass. Circular base with
three engraved scrolled feet, with
levelling screws, two marked 'A'.

Set into the base is a second plate with
inset silver compass surrounded with
double degree scale and two bubble
levels. This plate may be rotated
within the base, and has engraved on
its silvered surface a degree scale in
four quadrants and a table for
equation of time. Mounted on this
plate is a meridian ring engraved on
one side with a degree scale in four
quadrants and on the other with the
latitudes of various places. Within
this is an equinoctial engraved on the
face and inner edge with a 24-hour
scale and on the back with further
latitudes of towns. The bridge is
engraved with declination scales
within a formal stiff-oak leaf border.
Engraved on the edge of the
suspension ring mount, which is
adjustable for latitude, is a vernier
scale.

desired position. He would set the index to the appropriate declination on the bridge, and turn the dial, allowing it to hang freely, until a spot of light fell through the index onto the hour scale engraved on the second ring to indicate the time. If the index is correctly set this can only happen when the dial is hanging with the outer ring in the plane of the meridian (figs. 145, 146 and 147).

Dials of this kind may be found engraved with latitude scales for the northern hemisphere only or for both northern and southern. The back of the meridian ring often bears a 90° scale, the divisions being drawn radially from a small hole as centre on the opposite side of the ring. If a short brass pin is placed in this hole, and the suspension mount moved to zero position of the latitude scale, the diagram (known as the nautical ring) may be used for measuring solar altitude.

Crescent

A variant form of the equinoctial ring dial which is particularly associated with Johann Martin of Augsburg is the crescent dial (fig. 148). In this the hour ring is cut into two halves which are set back to back. The gnomon is a crescent shaped bar the tips of which lie in the axes of the semi-circular hour scale. This gnomon can be adjusted against a declination scale and the dial is self-orienting. Often beautifully made in silver and gilt brass by Martin, and by Johann Willebrand, the parts of the dial folded so that it might fit into a fish-skin or leather case. Later in the eighteenth century larger models designed for effect as striking ornaments were produced with adjustable feet and spirit levels. Like the equinoctial ring, the dial is self-orienting.

Cruciform

In the sixteenth century a small dial was sometimes made in the shape of a cross. A vertical dial was drawn on each side of the body of the cross, the arms forming the gnomons. If the cross was correctly oriented using the compass set in the base, and inclined in the plane of the equator, the time was shown on each dial. Dials of this kind are usually hollow and were perhaps used as reliquaries (fig. 150). Surviving examples rarely date from later than the first half of the seventeenth century, although James Ferguson gave instructions for drawing this kind of dial as late as the latter part of the eighteenth century.

Diptych

Perhaps the commonest form of multiple dial, made of wood, ivory or more rarely metal, is the diptych. In its simplest form it consists of two leaves hinged to open at right angles to each other. On the inner face of the lid is a vertical dial, on the lower, a horizontal dial with an inset compass. The dials share a common string gnomon. Dials of this kind were usually made of wood with the scales engraved or punched directly onto them. In the eighteenth and nineteenth centuries a number of such dials was produced in which the scales were printed or painted on paper and pasted onto the wooden leaves (fig. 153). Rarely early examples of this type are to be found made of brass or ivory.

The developed ivory diptych dial of the sixteenth and seventeenth centuries is, in striking contrast to the crudely made simple versions, one of the finest products of the diallist's art. In addition to the vertical and horizontal dials, there are usually several small pin-gnomon dials (often in the form of shallow scaphes). These may indicate Italian, Babylonian or Jewish hours, the declination of the sun, or the length of day and night. As well as these, polar and equinoctial dials were sometimes incorporated, and lunar volvelles, nocturnals, wind-roses and calendar scales might be included. Any free space was used to list the latitudes of places, or filled with decoration. Generally the engraving was filled with blue, red, green or black colouring.

Fig 148
German equinoctial dial, signed 'Johann Martin In Augspurg', c. 1700, gilt brass and silver.

Scroll-engraved base plate with two levelling screws and a 16-point compass rose in the centre. At one edge of the plate a pierced and engraved equinoctial plate is hinged with semi-circular hour scales on each side. Attached to the centre of this plate is a folding calendar scale along which slides the crescent-shaped gnomon. At the opposite side of the plate is a bracket carrying a plummet.

Fig 149
French crescent dial, signed on the
inclining crescent arc, 'Inventé Par
Jacques Baradelle A Paris', and on the
base, 'Baradelle A Paris Quay de
Lorloge du Palais', *c.* 1800, brass and
steel, size of base, 114 × 169 mm
(4½ × 6⅝ in.).

Rectangular base plate on which are
engraved the latitude of twenty-five
places, adjustable by three screw feet
and with rotatable bubble level
mounted at centre. Firmly hinged to
one end is the crescent-shaped hour
ring with acanthus leaf decoration
and hour scales on the sides. This hour
ring may be set against a folding
degree scale (0°–90° reading to ½°)
for latitude. Mounted on the centre of
the crescent is a hinged bridge with
calendar scale for declination and
carrying an adjustable double
gnomon.

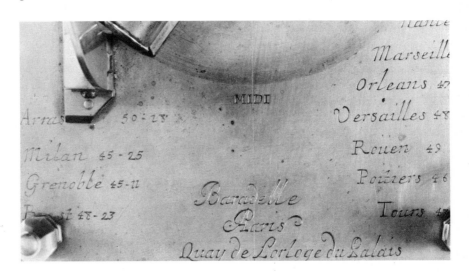

Fig 150
Crucifix dial, not signed, 1593, brass.

Hollow box in the shape of a cross
with suspension ring at the top; at one
end of the base, inside the dial, is a
compass in glazed box, also engraved
on the base is a scale of latitude
against which the body of the cross
can be set by a supporting strut.
Hour and zodiac lines are engraved on
the side of the instrument, and on the
faces are the latitudes of various
European towns. Although not
signed this dial is certainly the
product of the well-known Louvain
workshops.

Fig 151
German diptych dial, not signed, but
stamped '3' in two places, the
customary mark of Jacob Karner.
First half seventeenth century. Ivory,
with brass catches and fittings, the
engraving filled with red, green and
blue pigment. Length of longer side
92 mm (3⅝ in.).

Upper leaf outer face – 16-point
wind-rose, with the cardinal points
indicated in German and with brass-
tipped index rotating over a 32-point
scale numbered from the east. The
small hole is for viewing the north
point of the compass, and the compass
needle when the upper leaf is closed
down.

Upper leaf inner face – vertical pin-
gnomon dial for planetary hours and
the sun's declination. A table of
latitudes for sixteen places and fixing
holes for the string-gnomon at four
latitudes.

Lower leaf inner surface – horizontal
string-gnomon dial with hour lines
for latitudes 42, 45, 48 and 51; pin-
gnomon dials for Italian and
Babylonian hours. Inset compass
with magnetic declination mark and
the cardinal points indicated in Latin.
The base of the compass is stamped
'3'. Compass needle and glass
probably replacements.

Lower leaf outer surface – lunar
volvelle with table of epacts in
Julian and Gregorian calendars.
Decorated foliate borders
incorporating a second figure '3'.

A compartment chiselled in the side of
the lower leaf, with cover, is for
storing a wind vane, now missing.

Fig 152
German diptych dial, signed 'LIENHART MILLER 1605', 91 × 57 mm (3½ × 2¼ in.), ivory and brass.

Upper leaf outer face – 16-point wind-rose with the points named for use with a wind-vane carried in a small compartment chiselled in the side of the lower tablet.

Upper leaf inner face – vertical dial and vertical pin-gnomon dial.

Lower leaf inner face – horizontal dial surmounting inset compass, horizontal pin-gnomon dial for Italian and Babylonian hours.

Lower leaf outer face – lunar volvelle.

Fig 153
German diptych dial, signed on compass paper 'E. Ch. Stockert', eighteenth century, length 100 mm (4 in.), wood, brass and paper.

Two tablets of wood hinged to open at right angles have pasted to their inner faces papers printed with vertical and horizontal dial scales, and share a common string gnomon. A printed table of the latitude of fifty places is pasted to the lid, each of the printed papers has been hand decorated in water-colour. The string gnomon is set for latitude 50° (Cracow or Prague).

Fig 154

English compendium, not signed but probably by Henry Sutton, boxwood with brass latitude strut, catches and hinges, paper scale printed from an engraved plate, *c.* 1650, 121 × 123 mm (4¾ × 4⅞ in.).

On the exterior of the lid hinged to the lower and thicker leaf is an hour scale (1–12 × 2) divided in quarter hours, probably for a polar dial. On the interior is a vertical dial which shares a common string gnomon (missing) with the horizontal dial engraved on the base. Inset in the base is a blue-steel compass needle surrounded by a degree scale in four quadrants and mounted over a magnetic azimuth dial with 16-point compass rose at the centre. On the underside of the lower leaf (below) is a Gunter quadrant with a 90° scale along the two edges of the box and a shadow square. On a table engraved within the shadow square are the dates and times of rising of five stars which are marked on the projection. At the apex is a hole for a plumb line (missing) used in conjunction with the sighting pinnules (one missing) set in the side of the box.

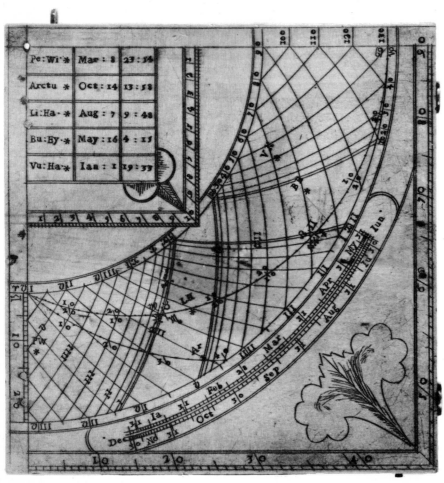

Astronomical compendia

Among the finest of all dials are the gilt brass instruments produced in the sixteenth and early seventeenth centuries which incorporate a number of different instruments. Square, round, octagonal or rectangular in shape and often very similar to the diptych dial, such instruments might incorporate a miniature astrolabe, an alidade, geographical map or nocturnal. Extremely beautifully produced, but expensive toys even in their own day, such instruments, judged by the number of signed examples that survive, seem to have been especially favoured by the German instrument-maker Christoph Schisler.

Polyhedral

A popular way of illustrating virtuosity as a sundial-maker in the sixteenth century was to design a dial to suit each face of a polygon, so that each dial would show the same time as the others. After falling somewhat out of favour in the seventeenth century, the form was revived when D. Beringer and other makers in eighteenth-century Germany began making cubical dials with coloured printed scales on each face (fig. 160). The chief use of the polyhedral dial was probably as a large fixed garden dial when it was cut in stone.

Simple azimuth

The simple azimuth is a horizontal dial engraved with a series of concentric hour rings, graduated in the solar azimuth angles for each month of the year. The curved hour lines are drawn across the corresponding azimuths, the 6 a.m./p.m. and 12 noon lines being straight since the sun is then due east, south or west. At the centre is a vertical pin index. When the dial is set with the 12 o'clock line on the meridian, the shadow of the index gives the time against the appropriate hour ring. This and the succeeding dial were often made with printed paper scales.

Magnetic azimuth

This dial resembles the simple azimuth but has a compass needle mounted on a pivot at the centre instead of the pin-gnomon. The four cardinal points are also usually marked and sometimes sights are added. In use the north point of the compass, or the sights, is directed towards the sun and the position of the compass needle against the appropriate hour ring indicates the time (fig. 161).

Bloud

A pleasing combination of the diptych and the magnetic azimuth dials was developed by Charles Bloud and other Dieppe instrument-makers in the second half of the seventeenth century. The dial is made of ivory and may incorporate polar, equinoctial and string gnomon dials like other diptychs. In the lower leaf is a compass with an elliptical hour ring (often of pewter) placed over the compass card, the index being replaced by the point of the compass needle. As the construction of the dial necessitates this being fixed, adjustment for declination is effected by moving the hour ring itself by a cam beneath the dial. The scale against which it is set is placed on the back. The dial is used in the same way as the magnetic azimuth dial. The diptych is turned until the shadow of the upper leaf falls exactly over the lower, and the time is then read off by the position of the compass needle (fig. 162).

Analemmatic

Thomas Tuttel designed a self-orienting dial and published it in 1698 in his *Description and Use of a New Contriv'd Eliptical Double Dial*. The instrument consists of two dials. One is an ordinary horizontal dial in which the time is measured by the sun's hour-arc (i.e. position in the equinoctial). The other dial is an azimuth dial formed from an orthographic projection of the equinoctial circle onto the plane of the horizon. An elliptical hour scale is thus produced with the 12 o'clock line on the minor axis. A

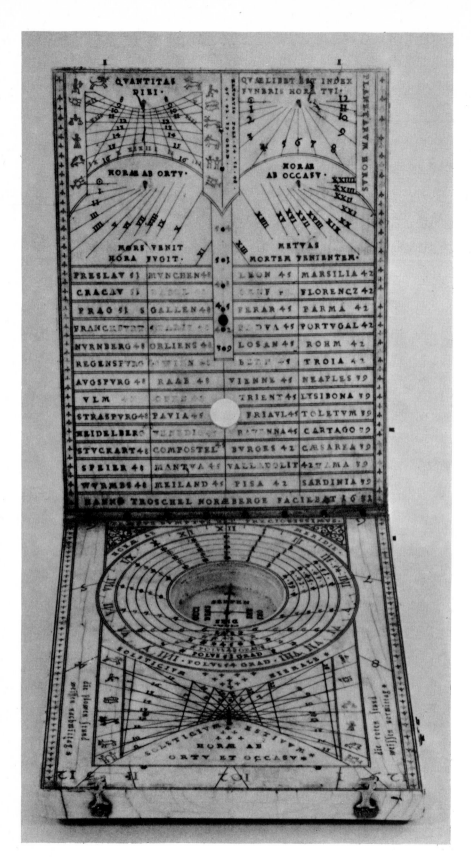

Fig 155
German diptych dial, signed 'HANS
TROSCHEL NORAE BERGE FACIEBAT 1631',
157 × 112 mm (6¼ × 4⅜ in.), ivory and
gilt brass.

Lower leaf outer face – table of
latitudes of a further twenty-eight
places, lunar volvelle with
aspectarium surrounded by
concentric tables for the hours of the
night, phase of the moon and the
epacts according the Gregorian and
Julian calendars.

Upper leaf inner face – four vertical
pin-gnomon dials for: the length of
day (marked '*Quantitas diei*'), equal
hours (marked '*Planetarum horas*')
and with the inscription '*Quaelibet est
index funeris hora tui*' (any hour is a
pointer to your death), Babylonian
hours ('*Hora ab ortu*'), Italian hours
('*Horae ab occasu*').

Beneath the third and fourth dials are
the exhortations '*Mors venit Hora
fugit*' (Death comes, the hour flies),
and '*Metuas mortem venientur*'
(Fear the coming of death).

In the centre are attachment holes for
the string gnomon at latitude 39°,
42°, 45°, 51°, 54°. The lower part of the
leaf contains a table of the latitudes of
fifty-two places.

Lower leaf inner face – horizontal dial
with hour scales for six latitudes
corresponding to those marked for
the six attachment holes in the upper
leaf. Inset compass, pin-gnomon dial
for Italian and Babylonian hours. At
the top is the inscription '*Temporis
sumptus est preciosissimus*' (Time's
consumption is most precious).

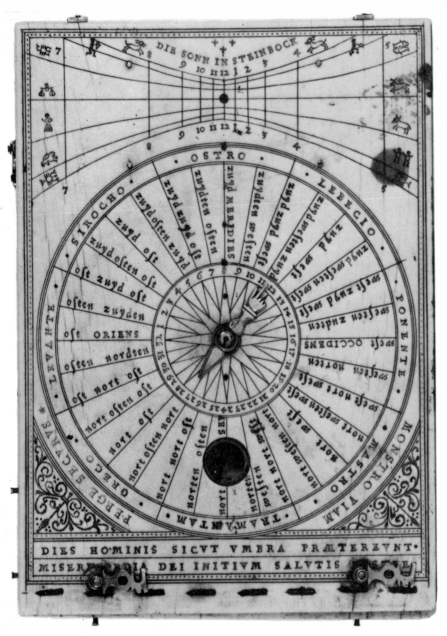

Upper leaf outer face – a wind-rose with 32 compass points named in German and the cardinal points named in German and Latin; rotatable gilt brass index; aperture for observing north point of the compass needle below; pin-gnomon dial for the sun's position in the zodiac. Beneath the wind-rose is the inscription '*Dies hominis sicvt vmbra praetereunt. Misericordia dei initium salvtis s E*' (Man's days have passed by unnoticed like the shadow. The pity of God is the beginning of salvation).

Fig 156
English compendium, signed 'C[harles] W[hitwell]', 1610, 80 × 45 mm (3⅛ × 1¾ in.). Gilt brass with paper compass scale printed from an engraved plate and coloured by hand. On the outer surface of the lid is a nocturnal with a zodiac/calendar scale, on the inside of the lid a table of primes and epacts.

Mounted on the same hinge as that joining the box and lid is an equinoctial dial which may be folded flat when not in use. A compass with double needle for orientation is contained in the base. On the outer surface of the base is a tide calculator.

Fig 157
German compendium signed 'Hans
Ducher zu Nurnberg', 1593,
diameter of major axis 60 mm (2⅜ in.),
gilt brass.

Outer surfaces of the compendium
shown in fig 124.

The lower lid is engraved with a wind-
rose in the form of a sunburst with a
rotatable index. The hole is for
observing the north point of the
compass set in the box below when
the instrument is closed. On the
lower lid is a rotatable volvelle and
scale for the length of the day.

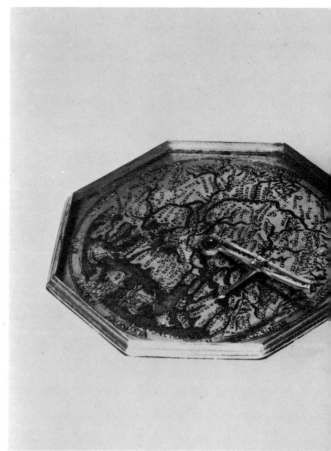

Fig 158
German astronomical compendium,
signed 'CHRISTOPHORVS SCHISLER ME
FECIT VINDELICORVM ANNO DOMINI', gilt
brass.

The compendium includes a
horizontal dial with five hour scales
for latitudes 42, 45, 48, 51, 54; two
geographical plates; lunar volvelle
and aspectarium; a Rojas dial with
calendar scale; wind vane; levelling
plummet and stand.

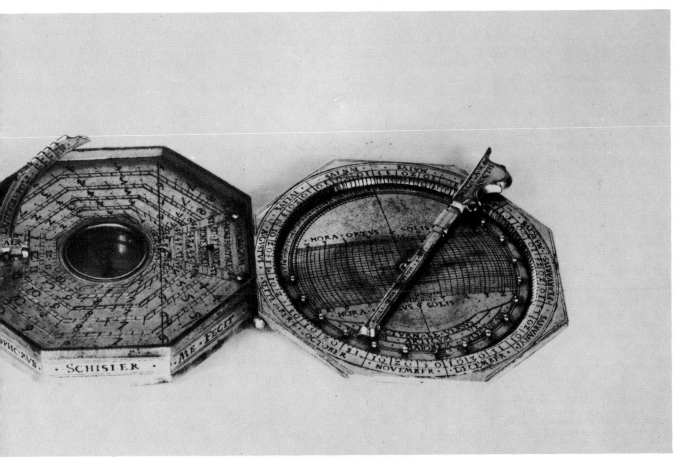

Fig 159
French horizontal sundial, mounted
on the inside back cover of a watch
signed 'Gribelin a Blois', c. 1600.

In the centre is a compass with
magnetic declination mark c. 11°
east, string gnomon with folding
support. Encircling the dial is the
legend '*Christe tu cordi polus entheo
quod spiritus magnes tvvs unxit
umbrae vita transcursus lare dum
sub atro clara micat mens*' (O
Christ, pole star to the divinely
inspired heart which the lodestone of
thy spirit has touched [anointed], life
is the passing of a shadow during
which the mind shines out brightly
from under its dark abode).

Fig 160
German polyhedral dial, signed
'D. Beringer', mid-eighteenth century.
Base 99 × 79 mm (3⅞ × 3⅛ in.), wood
with hand-coloured scales printed
from engraved plates; brass
gnomons.

Polyhedral dial in the form of a cube.
A dial is drawn on each face and when
the instrument is correctly oriented
using the compass set into the base, all
five dials will show the same time.

Fig 161
English magnetic azimuth dial and surveyor's compass, signed *'Henry Sutton fecit'* and dated 'Aprill 27 1650' (?6).

Brass with paper scale printed from an engraved plate; length of side of square 109 mm (4¼ in.), overall length 129 mm (5⅛ in.).

The outer printed scale is divided by tens in four quadrants (0°–90°/90°–0°). Within this, and surrounding the heart-shaped signature cartouche, are the hour lines drawn on a stereographic projection with a zodiacal calendar, In the centre is an 8-point compass-rose.

At a slightly later date the paper has been converted for use as a surveyor's compass. The circular scale having been mounted on a square paper, a second set of degree numbers (360°–0°) has been added outside the circle in a late-seventeenth or early-eighteenth-century hand. The initials of the cardinal points have been added in ink to the compass-rose in the centre, and the needle has probably also been replaced. The whole instrument has been mounted in a glazed brass box with extended back, having four holes and a three-quarter slot cut in the overhang, presumably for attachment to another instrument, such as a plane table.

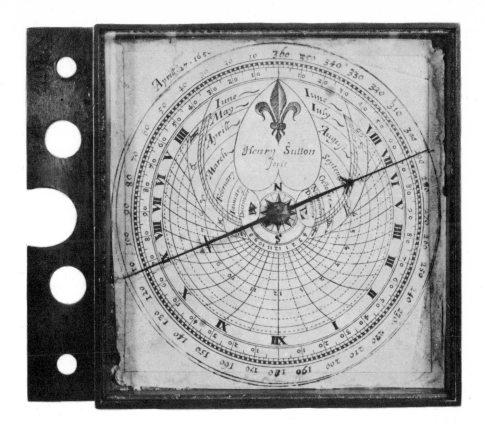

Fig 162
French magnetic azimuth diptych dial, signed 'Fait par Charles Bloud A Dieppe', late seventeenth century, length 100 mm (4 in.), ivory, pewter and brass.

On the outer face of the upper leaf is a horizontal dial and polar dial. On the inner face is a lunar volvelle and scale of degrees for latitude. In the base is a compass needle pivoted above a paper printed with the latitudes of various European towns, and an elliptical hour ring adjusted by setting the calendar scale on the underside of the base.

declination scale runs along this axis against which the vertical index is adjusted. In Tuttel's version the two dials are engraved on a brass plate with their 12 o'clock lines parallel (fig. 163). In other versions the dials may be superimposed upon each other (fig. 164). When the dial is placed horizontally and rotated the correct time is indicated when both dials agree.

Hour systems

Man's methods of reckoning time, even in Western Europe, have been many. Before the mid-fourteenth century the period of daylight was generally divided into twelve equal parts, and the period of night into twelve equal parts. Except at the equinoxes, the twelfth part of the day would not equal the twelfth part of the night, because the total length of day and night were not equal to each other. This system, which is occasionally found on sundials, is known as unequal, Jewish, or planetary hours. Equal hours, in which the whole period of day and night was divided into twenty-four equal parts, had long been used by astronomers, and are sometimes referred to as astronomical hours. With the development of the mechanical clock which could measure no other form of time, equal hours gradually supplanted the unequal systems. Two other hour-reckoning systems which are occasionally found on Renaissance sundials are those of Babylonian hours *(horae ab ortu solis)*, equal hours counted up to twenty-four from half an hour after sunrise, and Italian hours *(horae ab occasu solis)*, equal hours counted from half an hour after sunset. Each of these systems is a twenty-four hour system.

By the end of the seventeenth century a greater uniformity was making its way among the hour systems of Europe, and the ingenuity of makers, shows the influence of mechanical time-keepers, both in the mechanical equinoctial dial and the watch-cased compass dials. However, in the absence of a general and uniform time standard clock and watch owners needed sundials for checking and setting. It was not the clock, but the coming of the electric telegraph and the development of national and later universal time services, which removed the sundial from its primary position as the only reliable time-finding instrument. With its passing from common use in the middle years of the nineteenth century one of the major vehicles for popular understanding of astronomy and geometry disappeared.

Fig 163
English analemmatic dial, signed 'Tho. Tuttell Charing X Londini Fecit', c. 1700, length open 158 mm (6¼ in.), brass.

Two plates hinged to lie flat when open. On the right-hand plate is a horizontal dial with solid folding gnomon and 8-point compass-rose and a narrow border of stiff oak leaf decoration. On the left-hand plate in the illustration is an analemmatic dial with calendar scale in the centre against which the gnomon is set. Around the edges of three sides of the plate is a protractor scale. On the reverse is a perpetual almanac with scales for times of sunrise and sunset.

Fig 164
English analemmatic dial, signed
'T. W. PARRY, OPTICIAN, 24, Holywell St.,
STRAND, LONDON', *c.* 1845, 155 mm
(6⅛ in.), brass.

Circular base plate with three screw-
adjustable feet and a spirit level.
Mounted behind the level is a curved
brass bar which slots into an eye on
the dial plate which may be inclined
for latitude. The dial plate is
composed of two sections: an
elliptically shaped portion, engraved
round the edge with the analemma
hour scale, and in the centre with a
zodiac/calendar scale, and a square
plate engraved with a horizontal
dial hour scale with triangular
gnomon. In the centre is an equation-
of-time scale.

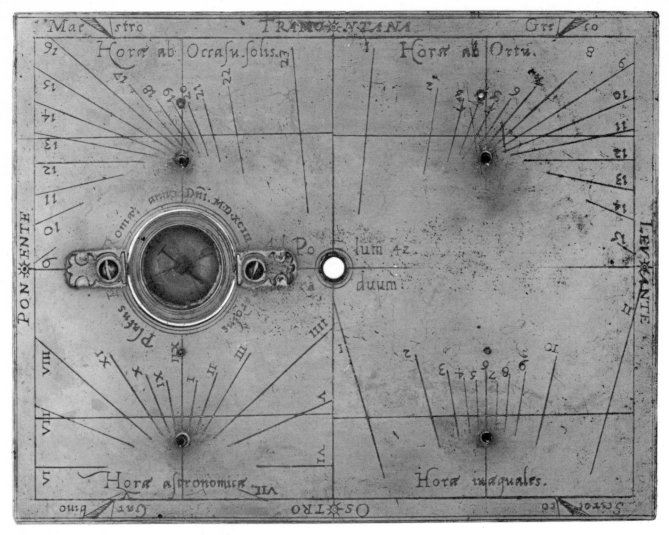

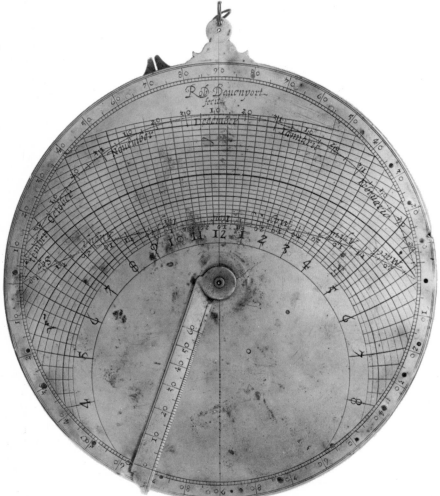

Fig 165
Italian pin-gnomon horizontal dial,
signed 'Carolus Platus F. Romae anno
Dni. MDXCIII' (1593), 157 × 120 mm
(6¼ × 4¾ in.).

Rectangular plate with four
horizontal pin-gnomon dials on each
side; inset compass engraved for use
in latitude 42° the other for latitude
48°, and each carries four dials for
Italian, Babylonian, equal and unequal
hours. Round the edges of the plate
are the names of the four cardinal
points and various units. The central
hole was presumably used for
supporting the plate when in use.

Fig 166
English/Scottish horizontal dial,
signed 'Rob Davenport fecit', c. 1650,
brass, diameter 152 mm (6 in.).

Circular disc engraved round the edge
with a degree scale in four quadrants.
Within this is an example of William
Oughtred's double horizontal dial,
invented c. 1620–30, which employs a
stereographic projection as in an
astrolabe. At the centre of the plate is
a graduated rule.

Fig 167

German mechanical equinoctial dial, signed 'Hahn', c. 1770, brass and steel.

Heavy meridian ring graduated in one quadrant with a fixed steel rod in the polar axis. A rectangular frame revolves on this rod, and has a plate at each end engraved with the signs of the zodiac, and with an equation-of-time scale. In one of the plates are two pinholes. The frame is geared to drive the hands attached to the watch face between the two plates. In use the frame is rotated until the two spots of light fall on the scale of zodiacal signs appropriate to the time of year and the correct time is shown by the watch dial.

Phillip Matthias Hahn (1739–1790) was a clock and watchmaker at Ostmettingham who supplied dials of this kind, developed from the universal equinoctial ring dial, with his clocks so as to check their accuracy.

Fig 168
Flemish horizontal dial, not signed,
seventeenth century, overall length
78 mm (3 in.), brass, ivory with amber
filling in the roundels.

Five part barrel engraved with
flowing foliage designs and with
pierced roundels on the lid. The upper
portion of the body contains an 8-point
compass-rose on paper printed from
an engraved plate and coloured by
hand. An hour scale and folding
gnomon (now missing) may have sat
above the compass or in one of the
other compartments. The second
section contains a lunar volvelle, the
remaining compartment which is
empty may have acted as a snuff box.

4

Surveying

A GEOMETRICAL PRACTICAL
TREATIZE NAMED PANTOMETRIA,
diuided into three Bookes, LONGIMETRA, PLANIMETRA, and
STEREOMETRIA, Containing rules manifolde for mensuration of all *Lines*,
Superficies and *Solides*: with sundrie strange conclusions both by Instrument and with-
out, and also by *Glasses* to set forth the true Description or exact Platte of an whole
Region. First published by *Thomas Digges* Esquire, and Dedicated to the Graue,
Wise, and Honourable, Sir *Nicholas Bacon* Knight, Lord Keeper of the great
Seale of England. With a Mathematicall discourse of the fiue regular
Platonicall Solides, and their *Metamorphosis* into other fiue compound rare *Geometricall Bodies*, conteyning an hun-
dred newe *Theoremes* at least of his owne *In-
uention*, neuer mentioned before
by anye other *Geome-
trician*.

LATELY REVIEWED BY THE AvTHOR
himselfe, and augmented with sundrie *Additions, Diffini-
tions, Problemes* and rare *Theoremes*, to open the passage, and prepare away to the vnderstanding
of his Treatize of *Martiall Pyrotechnie*
and great *Artillerie*, hereafter to
be published.

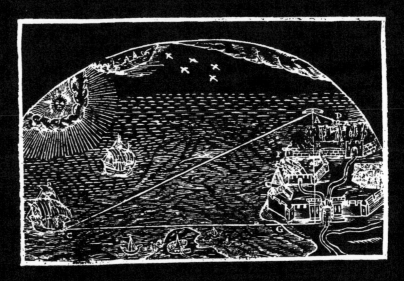

AT LONDON
Printed by *Abell Jeffes*.
ANNO. 1591.

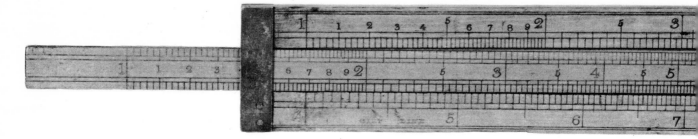

'G EOMETRY originally signified the art of measuring the earth, or any distance or dimensions on or within it; but it is now used for the science of quantity, extension, or magnitude, abstractly considered, without any regard to matter,' wrote a mathematics teacher, William Davis, in 1798. Geometry and trigonometry lie at the basis of the science of land surveying, both geodetic and topographical. This, however, was hardly appreciated before the sixteenth and seventeenth centuries. Land surveyors before then measured land with only the rudiments of mathematics, and less accuracy. Topographical surveying with which this chapter is primarily concerned was the Cinderella science until land became expensive. The surveyor only became important when fields and estates were enclosed and registered for taxation.

Surveying instruments constitute some of the most spectacular three-dimensional sculptures available to the collector. The principles of surveying are contained in the modern tachymetric theodolite which can simultaneously fix the three polar co-ordinates of altitude, azimuth (compass bearing) and distance, but the complex refinements to achieve this end developed slowly. Although the theodolite had become a reasonably accurate instrument capable of measuring to within 30 minutes of arc, or less, by the beginning of the eighteenth century, most surveyors could not afford one, for the fees they commanded for small enclosures were so miserable that they had little to spend on sophisticated equipment.

The theory of surveying has not changed since the ancients, although in practice complex instrumentation has speeded things up. The Egyptians, adept at land measuring and conversant with geometry, placed three men in a straight line with their forefingers raised as markers. Two of them were quite close together, but the third was placed at the point to be considered. This gave them a line of sight, which could be staked out and measured. This is also the principle of the alidade, which is a movable pointer with sights in the upturned ends, through which a desired location can be viewed. Familiar at first on the astrolabe, probably known to the Greeks, like many other discoveries it found its way into surveying.

The problem of accurately measuring land was solved by triangulation, that is, by dividing the area, of whatever geometrical shape, into triangles, measuring each one and computing the total.

Discussed by Gemma Frisius (1508–1555) in 1529, one of his nephews Gualterius Arsenius was the first to construct an astrolabe with an inset compass, to make it more suitable for surveying. The shadow square on the reverse of the mater, was used for centuries to determine altitudes for it was a rough scale of the tangents and co-tangents of angles (figs. 10, 12, 14 and 170–73). With the alidade and the compass it was possible to line up azimuths, while distance was measured with cords, rods or chains.

Rod, rood, pole or perch were Saxon terms to represent area as well as length. Before metrication is completed, some of us may still remember the incantations of our schooldays. Rods, etc., were of different lengths to measure specific types of land, i.e. farmland, grazing, woodland, etc., and termed *customary* rods, and the area so measured was termed *customary* acres. Acre, from the German *Acker*, meaning field or arable land, was a surface area which could be ploughed by a yoke of oxen in one day, and was the size of 160 square rods, regardless of the length of the rods. Weights and measures were not standardized in Britain until the reign of Queen

Fig 169
English timber measuring ruler (detail), not signed, *c.* 1870, length 610 mm (24 in.), boxwood, with brass caps.

Sliding timber rule with girt and inverted line scales.

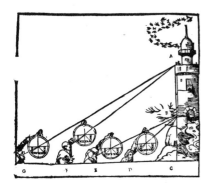

Fig 170
Plate from Ioannes Stoflerus *Elucidatio fabricae ususq. astrolabii* Paris 1618, p. 168.

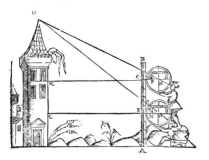

Fig 171
Measuring the height of a tower with the shadow square on the back of an astrolabe, from Cosimo Bartoli *del Modo di misurare* ... Venice, 1589, f36v.

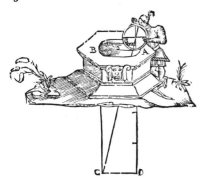

Fig 172
Measuring the depth of a well with the shadow square on the back of an astrolabe, from Cosimo Bartoli *del Modo di misurare* ... Venice, 1589, f46r.

Fig 173
Using an astrolabe to measure the angle of inclination for pieces of field artillery.

From Cosimo Bartoli *del Modo di misurare* ... Venice, 1589, f53v.

Elizabeth I, although the first attempt at a statutory policy had been introduced in 1277 during the reign of Edward I (1272–1307). The *statutory* rod was established at 16½ feet based on a German definition, and the *statutory* acre measured 160 square *statutory* rods.

Modern surveying began in the sixteenth century when the level of land values rose due to a general improvement in economic conditions. Land was enclosed, and owner's rights for each lot were verified with the Land Registry Office, termed the Cadastre. Thus cadastral surveying made more demands on the land surveyor, who, in his turn, demanded precision instruments suitable for his profession from the skilful craftsmen who were well established by this time. Cartographers were the local land surveyors, and by 1780 every county in England had been surveyed and mapped on the scale of one inch to a mile.

The early surveyor, signifying overseer from the French, was responsible to the lord of the manor to appraise his land by 'viewing' from a vantage point, which is in fact what many of them did, for without suitable equipment and definite training for 'plotting the dimensions' they had to assess the size of an area by guesswork.

Books on surveying in English started with an anonymous volume in 1523 which advised its readers to use a cord or chain with a 'Dyall' on cloudy days to determine the cardinal points. It was not made clear what this instrument actually was, but it was probably a pocket sundial of some sorts with an inset compass.

A BOOKE
named Tectonicon,

briefely fhewing the exact meafuring, and fpee-
die reckoning all maner of Land, Squares,
Timber, Stone, Steeples, Pillers, Globes, &c. further,
declaring the perfect making and large vfe of the Carpenters Ruler, containing a Quadrant Geometricall: comprehending alfo the rare vfe of
the Squire. And in the end a little Treatife adioyned, opening
the compofition and appliancie of an Inftrument called
the profitable Staffe. With other thinges plea-
faunt and neceffary, moft conducible for
Surueyers, Landmeaters, Ioyners,
Carpenters, and Mafons:

Publifhed by Leonard Digges
Gentleman, in the yeare of our
LORD. 1556.

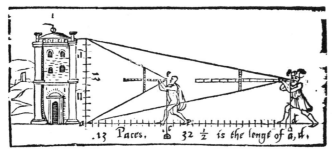

.13 Paces, 32 ½ is the lengt of a, H.

Imprinted at London in Fleete-
ftreate, neere to S. Dunftanes
Church, by Thomas
Marfh.

1585.

Fig 174
English pantograph, signed 'G. Adams London', c. 1770, overall length (closed) 508 mm. (20 in.), brass with ivory wheels.

Four brass arms rigidly attached to each other. On the centre arm marked D is a scale of proportions from 1:12 to 1:½ against which a heavy weight may be set by a slider to supply the central pivot of the instrument. A corresponding scale is marked on the pointer arm B. Pantographs of this kind were commonly included among eighteenth-century surveyors' equipment for producing enlarged or reduced copies of estate plans.

Fig 175
Title page of Leonard Digges *A Book named Tectonicon . . .* 1585. First published in 1556 by the engraver and instrument-maker Thomas Gemini, Digges's book was the first general introduction in English to geometrical surveying and measuring. He illustrates a cross staff being used to find the height of a tower, and the book described its use in considerable detail.

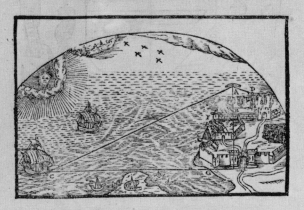

Fig 176
Title page of Leonard Digges *A
geometrical Treatise named
Pantometria* ... 1591.

First published posthumously in 1571
by Digges's son Thomas with some
enlargements of his own, *Pantometria*
provided a full introduction to
surveying and range-finding. The first
part, *Longimetria*, contained a
description of the 'topographical
instrument' (see fig. 177) one of the
earliest forms of theodolite.

A GEOMETRICAL PRACTICAL
TREATIZE NAMED PANTOMETRIA,
diuided into three Bookes, LONGIMETRA, PLANIMETRA, and
STEREOMETRIA, Containing rules manifolde for mensuration of all *Lines*,
Superficies and *Solides*: with sundrie strange conclusions both by Instrument and with-
out, and also by *Glasses* to set forth the true Description or exact Platte of an whole
Region . First published by *Thomas Digges* Esquire, and Dedicated to the Graue,
Wise, and Honourable, Sir *Nicholas Bacon* Knight, Lord Keeper of the great
Seale of England. With a Mathematicall discourse of the fiue regular
Platonicall Solides , and their *Metamorphosis* into other fiue compound rare *Geometricall Bodyes* , conteyning an hun-
dred newe *Theoremes* at least of his owne *In-
uention*, neuer mentioned before
by anye other *Geome-
trician*.

LATELY REVIEWED BY THE Avthor
himselfe, and augmented with sundrie *Additions, Diffini-
tions, Problemes* and rare *Theoremes*, to open the passage, and prepare away to the vnderstanding
of his Treatize of *Martiall Pyrotechnie*
and great *Artillerie*, hereafter to
be published.

AT LONDON
Printed by *Abell Jeffes*.
ANNO. 1591.

The development of the theodolite

The term 'theodolite' has been ascribed to many different forms of surveying instruments which, while varying in structure, contrive by various
means, to arrive at the same solutions.

The first real attempt to educate the surveyor was in a book by Richard
Benese (d. 1546) published in 1537, in which he explained how plotting
dimensions should be carried out in a practical manner. He explained
how irregular areas should be divided into triangles and how to compute
them, and how the surface area of hills and valleys should be calculated,
how acres were always measured width by breadth and advised the
carrying of a notched stick as an *aide memoire*. Like the mariner, the
surveyor was an ignorant but a practical fellow. Resistance to change
retarded progress, and the influence of new techniques and even overtly
popular textbooks like Leonard Digges's *Tectonicon* of 1556 only per-
colated through gradually.

Leonard Digges was one of the outstanding British technical writers of
the sixteenth century. He believed that the artisan should be introduced to
elementary mathematics, and his political activities which caused him to
join Wyatt's rebellion, ruined his career.

Through his friendship with Dr John Dee (1527–1608), the celebrated
Elizabethan scholar, he became acquainted with the works of Peter Apian,
Oronce Fine and Gerhard Mercator, who were Dee's closest friends. This
select coterie of sixteenth-century master-minds bore a great influence on
Digges, who drew heavily on Apian's *Instrument Buch* to describe the three

Fig 177
English theodolite, signed 'H. Cole',
1586, brass, 180 mm (11¼ in.). Vertical
semi-circle divided in two quadrants
from the centre point. Within this is a
shadow square and along the
horizontal edge is the sighting arm.
Below the scale of degrees and above
the compass hangs a levelling
plummet. The horizontal circle is
divided into 360° with 32 compass
directions marked. Within this circle
is a geometrical square. The compass
needle and glass, and three sights for
the horizontal circle, are missing.

One of the earliest known theodolites,
the instrument is comparable with
that described and illustrated in
Leonard Digges *Pantometria* 1571 (see
fig. 176).

Fig 178
German pedometer, engraving signed
'\\&', late sixteenth century, diameter
82·5 mm (3¼ in.), gilt brass and steel.

Circular brass box, steel back with
steel clips. On the dial are two
concentric scales for counting paces;
the outer numbered from 5–100/0; the
inner 1000–1200/0. Finely engraved in
the centre is a prospect of a walled
town seen from a distance. In the
foreground and to the left, trees and
bushes are growing while behind them
and to the right, the towers and spires
of the town are to be seen. A bird
(a stork) is seen roosting on the top of the
tallest building, whence a second bird
seems to have just flown. Behind the
town is a mountain peak behind which
the sun appears, either rising or
setting.

Brass hands, of which the upper (for
the outer scale) is a replacement.

Projecting through the side of the case
from the recording mechanism inside
is a steel eye. In use the instrument
was clipped to the wearer's belt, and
a string attached to the eye was tied
either to the knee or to the foot. At
each step taken, therefore, the eye
would be jerked down, thus causing
the hands to be advanced one division.

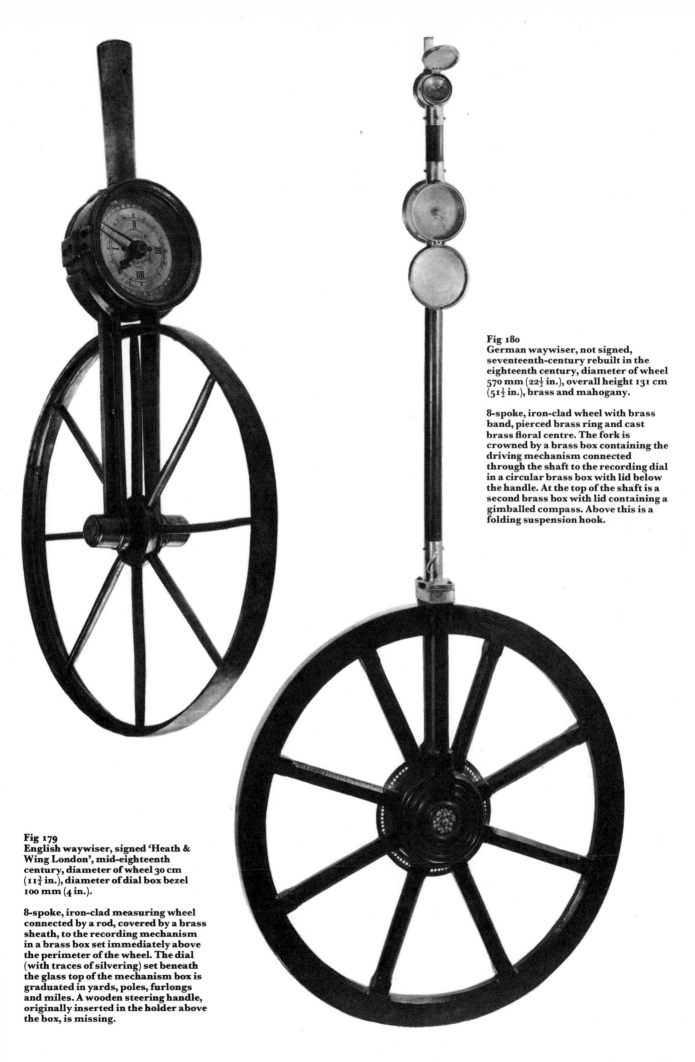

Fig 180
German waywiser, not signed,
seventeenth-century rebuilt in the
eighteenth century, diameter of wheel
570 mm (22½ in.), overall height 131 cm
(51½ in.), brass and mahogany.

8-spoke, iron-clad wheel with brass
band, pierced brass ring and cast
brass floral centre. The fork is
crowned by a brass box containing the
driving mechanism connected
through the shaft to the recording dial
in a circular brass box with lid below
the handle. At the top of the shaft is a
second brass box with lid containing a
gimballed compass. Above this is a
folding suspension hook.

Fig 179
English waywiser, signed 'Heath &
Wing London', mid-eighteenth
century, diameter of wheel 30 cm
(11¾ in.), diameter of dial box bezel
100 mm (4 in.).

8-spoke, iron-clad measuring wheel
connected by a rod, covered by a brass
sheath, to the recording mechanism
in a brass box set immediately above
the perimeter of the wheel. The dial
(with traces of silvering) set beneath
the glass top of the mechanism box is
graduated in yards, poles, furlongs
and miles. A wooden steering handle,
originally inserted in the holder above
the box, is missing.

indispensable instruments for the surveyor in *Tectonicon*. These were: the geometrical square borrowed from the astrolabe, engraved onto a wider rule, and used with a plummet, it served as a graphical means to roughly determine the tangents and co-tangents of angles; the carpenter's square; and the cross staff (Digges called it a 'Profitable Staff').

He described two new instruments in *Pantometria*. These were the theodolitus, which is not the one we now consider a prototype theodolite, and a more advanced instrument, the topographical, which was a composite instrument which embodied all the known surveying principles at that time. With it, angles in the horizontal plane and elevation could be determined.

It consisted of three parts: first, the geometrical square, which was a horizontal limb on which (second) the circle of the theodolite was constructed, and (third) a semi-circle attached to the horizontal limb at the centre of the theodolite. Humphrey Cole (1530–91) made some instruments after Digges's designs of which two extant are known, one belonging to St John's College, Oxford, and the other in the National Maritime Museum, Greenwich (fig. 177).

Early measuring instruments

While technical reforms were proposed the surveyor continued to do his best with whatever methods he could improvise. Like the mariner, his rule of thumb assessments were adequate for the approximate definitions required at that time.

The first record of the instruments he carried was described and drawn by Cyprian Lucar in 1590: the 4 perch line, a length of wire divided with a carpenter's rule into four perch lengths, then into half and quarter perches, then into foot lengths, each painted in distinguishing colours; the plane table, consisting of a rectangular board on which to place the drawing paper and a folding frame which gripped the edges to hold it in place, plus a ruler or index of some kind, and the tripod to hold the plane table, which evolved from a single stake that became a solid foundation by the addition of three short feet near the base, capable of lining up the plane table with the horizon.

Scholars continued to contribute to the science of surveying, and Edmund Gunter (1581–1626) professor of astronomy at Gresham College 1619–26, already mentioned in connection with his mathematical aids for the mariner, proposed an innovation, which still bears his name. As an improvement on the 4 perch line and Rathborne's chain, he produced the Gunter chain, which, he suggested, should consist of 100 links, marked off with a brass ring at every tenth, the total length being 66 feet. To be used with the statutory rod as a standard of land measurement, every statutory acre of 160 square rods measured with a Gunter chain consisted of 100,000 square links, and any area in square chains could be reduced to acres, roods and square rods. The advantage of the chain was that it could be used from either end.

Linear measurement was originally related to the human body, thus the terms were only approximate lengths. Standard measurement came much later, with the work of Picard and the French Academicians in the late seventeenth century and George Graham's standard yard at the beginning of the eighteenth century. The accurately graduated steel measuring tape with which we are familiar was introduced in 1905. Paces were a favourite form of measurement, used in antiquity and by the Romans who called the distance covered in a thousand paces one *milla*, approximately 1609 metres (a Roman pace being a stride of both the left and the right legs). Several sixteenth-century examples survive of an instrument, the pedometer, which recorded this kind of humanly measured distance. Attached to the wearer's belt, this was connected to either the knee or the foot by a string looped through a metal eye, so that at each step taken the eye would be agitated and cause a pointer on a scale to move a division. An inner scale computed the number of paces to record the distance covered (fig. 178).

Fig 181
English waywiser, signed 'W. & S. Jones, 30 Holborn, LONDON', c. 1860, diameter of wheel 80 cm (32 in.), diameter of dial 180 mm (7¼ in.), mahogany and brass.

6-spoke, iron-clad wheel connected from the axle by a cased rod to the recording mechanism in the handle-arm. Silvered brass dial graduated in yards, poles, furlongs and miles; blued-steel hands, glazed lid with brass hinges signed 'MOORE & CO. PATENT'. A small compartment in the handle arm contains a brass spanner.

Paul Moore took out a patent for the manufacture of hinges (No. 434) on 4 March 1858.

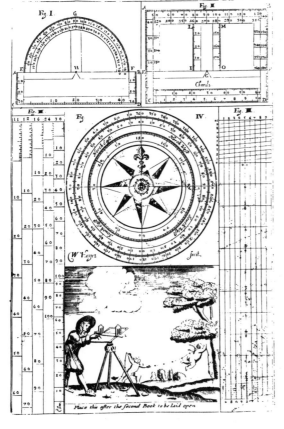

Fig 182
English plane table ruler, not signed,
c. 1700, 317 × 389 mm (12½ × 15¼ in.),
boxwood with brass hinges.

On the inner edge of each arm (both
sides) is a scale of inches divided in
tenths, on the outer edges is a double
degree scale with the divisions radial
from the centre.

Centre
Italian sighting compass, not signed,
seventeenth century, diameter
140 mm (5½ in.), brass.
Circular brass plate with attachment
holes, divided round the edge in four
quadrants 0°–90°/90°–0°/0°–90°/
90°–0°, and with eight wind directions
marked. An index with triangular
base rotates above this plate, and
above this is a circular disc with two
sights. Mounted in the centre is a
compass in glazed box with the lid
engraved with a coat of arms and the
slogan 'LAUS D [EO]' (? later).

Fig 183
Engraving from William Leybourne *The
Compleat Surveyor . . .* fifth edition
with additions by Samuel Cunn,
London, 1722. Leybourne was one of
the surveyors of London after the
great fire in 1666.

Fig 184
Measuring horizontal angular
distance with an astrolabe, from
Cosimo Bartoli *del Modo di
misurare* ... Venice, 1589, f5or.

Fig 185
Italian (?) clinometer, not signed,
seventeenth century, brass.

Triangular plate cut away to leave a
semi-circular arc divided from the
centre point 0–12 and a rectangle
divided from the mid-point of the
longer side 0–6–0 in each direction.
The numbered divisions of the
rectangle are radial from a point at
the centre of its undivided side with
the half divisions of each of the
numbered points on the semi-circular
scale. A plummet/index (now missing)
was suspended from the small hole at
the apex of the instrument. The two
larger holes were for attachment to
some other and larger instrument.

G 2   *aiuto*

Fig 186
Dutch circumferentor and altitude-measuring device (Holland circle), signed 'C [&] D Metz Amstelodami', c. 1700, brass.

Circular cut-away plate with four fixed sights, a suspension ring being attached to the mounting of one of them, engraved round the edge with a double degree scale 0–180/180–0 and 0–360. Inside this in one quadrant is a scale 0°–90°–0° marked 'umbra versa, umbra recta' which functions as a shadow square. In two other quadrants is a scale for regular polygons divided 3–22. Rotating over this plate is a second, smaller, plate carrying the movable sights and with an early form of vernier. Mounted at the centre of the instrument is a compass with 8-point compass-rose. The north point is marked with a *fleur de lys,* the remaining cardinal points with initial letters. Set above this, with the hour divisions marked round the edge of the compass box, is an horizontal sundial. As shown in the photograph, the instrument is suspended for taking an altitude observation. When used as a circumferentor, it was mounted on a universal staff-head on a single adjustable column (see fig. 219). In addition a telescope could be fitted to the movable sighting and extended sights added to one pair of the fixed sights.

A finely made instrument which, in its double function, stands towards the end of the line of development of the circumferentor which began with the scales customarily placed on the back of an astrolabe.

The hodometer, waywiser or perambulator, was known in principle to the Greeks and Romans, but was re-invented in the seventeenth century to measure linear distances, and was a useful tool to plot road maps. Elizabeth Longford records in her biography that Wellington insisted that a perambulator be used to register the distances his army marched throughout his campaign in the Peninsular War (figs. 179–181).

The plane table was used with chains or rods for triangulation of small enclosures and estates. In *The Compleat Surveyor* William Leybourn (1626–1716) suggested that the folding frame used to clamp the chart to the plane table should be graduated with degrees of a circle, so that the bearing could be established without expensive instruments (fig. 182).

His book, published in 1653, ran to four editions in the seventeenth century and a revised edition was published by Samuel Cunn in 1727 (fig. 183) which was used for several generations. In his book, Leybourn discussed the theodolite and the circumferentor, which he recommended for large surveys.

The circumferentor, which should be distinguished from the earlier Holland circle, consisted of a compass surrounded by a graduated brass ring, divided into 360° along which two sight rules could be manoeuvred. These sight rules, otherwise known as alidades, stood vertical to the outer ring and had two slots cut longitudinally and parallel to the edge of the strip. (These sights later bore a fine wire through the centres.) Setting the needle to north, the sight rules could be fixed on two pre-determined points, so that their azimuths could be recorded, and the survey be related to cadastral maps (figs. 184, 186 and 187). It became apparent that there was no necessity for a complete circle as a semi-circle or even a quadrant was sufficient to take land-surveying sights.

An account of an instrument termed a *graphometre* was published in 1597 by Phillipe Danfrie, together with a protractor with a long jointed arm pivoting from its centre with which to transfer the physical findings to paper. The graphometer (to give its English spelling) consisted of a semi-circular brass plate, graduated in degrees on its outer limb, mounted onto a jointed knuckle, so that the instrument could be used to measure vertical or horizontal angles. When used in the vertical plane a plumb line would fall over the scale to convert the instrument into a clinometer (see p. 162 and fig. 185). Along the diameter, a fixed alidade extended well beyond the semi-circle, and another alidade of equal length pivoted from the centre, underneath the compass which was mounted on the axis (figs. 190 and 189).

Later on, the alidades were shortened (fig. 191) and later sometimes replaced with telescopes, but the instrument retained its popularity until the end of the eighteenth century. The graphometer was a French instrument, and most examples that have survived are by French makers. They embody the style and craftsmanship usually associated with French workshops and the decorative flourishes of the engraver are in evidence to add elegance to science (fig. 190).

The quadrant with telescopes and spirit levels was also used in surveying and the handsome tooled leather cases stamped with *fleur de lys* provided with some French instruments indicate that they were probably employed by civil servants.

A specially devised geodetic quadrant (fig. 193), which could also be used for astronomy, was based on the large mural quadrants designed for observatories. The problems of casting these large quadrants in one piece were enormous, so in practice they were constructed of sections bolted together, but an instrument whose radius is only 320 mm (12⅝ in.) was free from such difficulties. The whole frame with strengthening struts was cast in one piece of brass alloy, bearing all the refinements known at the time. Geodetic quadrants of this type were also made in wood, with several stretchers for extra strength, but the metal frames proved more rigid and subsequently more accurate.

Surveying had gained considerably in importance by the end of the seventeenth century when new tracts of land in America were being

Fig 187
English circumferentor, signed
'T. Heath Fecit', first half eighteenth
century, diameter 230 mm (9 in.),
brass.

Silvered compass with inset spirit-level, engraved at the circumference with two degree scales, the inner in four quadrants, the outer 0°–360°, mounted in a glazed box, on a cross-frame, with two detachable sights. This frame rotates over a circular cut away plate engraved on the circumference with a 0°–360° degree scale against which the fiducial edge of the cross-piece is read. Two further detachable sights are fixed to the plate which is mounted on a universal staff-head. On either side of the compass box are two wing nuts to which a separate bridge may be attached to carry further sighting devices.

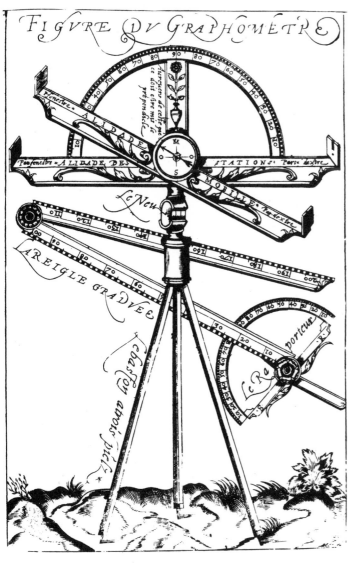

FIGVRE DV GRAPHOMETRE

Fig 188
French graphometer, engraving on
paper from, Philippe Danfrie
*Declaration de l'usage du
graphometre* ... Paris, 1597, p9.
Compare Danfrie's instrument
illustrated in fig. 190.

Fig 189
French graphometer, signed 'Dubois
a Paris', eighteenth century, overall
width 352 mm (13⅞ in.), brass and
steel.

Pierced semi-circular plate with
centrally placed silvered compass,
engraved on the raised circumference
with a 360° scale, mounted in brass
box, supported from below by a
square plate suspended by four
screws. On the semi-circular edge of
the instrument is a double degree
scale (o–180/180–o) having two fixed
sights at its ends with the sighting
vanes of catgut aligned on the o/180
mark. Pivoted at the centre, and freely
rotatable, is an alidade with fine scales
on the fiducial edges. The plate and
alidade are engraved with restrained
foliate decoration, and mounted on a
staff-head by a ball-joint adjustable
by a wing-headed screw.

Fig 190
French graphometer, signed
'P. Danfrie', *c*. 1597, brass.

Semi-circular ring with extended and
thickened horizontal member marked
'*Alidade des stations*' and with slit
sights. The semi-circular arc,
strengthened by a decoratively
engraved single strut, is engraved
with a degree scale o°–90–o. Attached
at the centre point and rotating over
the scale is a second alidade marked
'*Alidade mobille*'.

The development of the graphometer
from the sighting alidade and
quadrantal degree scale can be clearly
seen in Danfrie's innovatory
instrument which should be
compared with that shown in fig. 188.

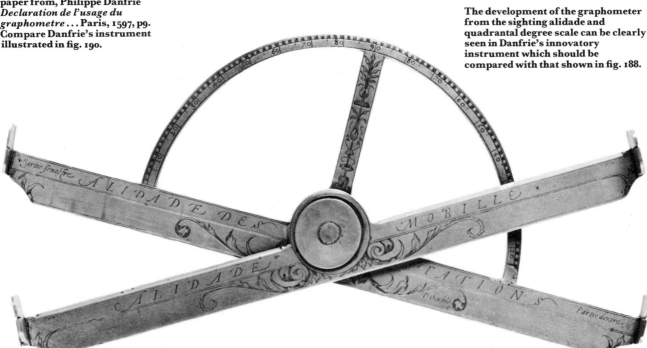

Fig 191   French graphometer, signed 'Canivet A Paris', c. 1760, brass.

Pierced semi-circular plate, carrying at the centre a silvered compass with a degree scale engraved on its raised circumference; surrounding the rim of the compass is a border of fine-leaf decoration. On the semi-circular edge of the instrument is a double degree scale (0–180/180–0) having two fixed sights at its ends. Pivoted at the centre, and freely rotatable, is an alidade. The whole instrument may be mounted on a staff-head by a universal ball-joint.

Fig 192   French graphometer and water level.

Graphometer signed 'Lennel Elêve et Successeur de M' Canivet à la Sphère à Paris 1774', 1774, length of side 324 mm (12¾ in.), length of telescope 386 mm (15⅛ in.), brass.

Pierced semi-circular plate, with restrained foliate decoration, carrying at the centre a silvered compass with degree scale (0–360) on its raised circumference. This compass is contained in a glazed box suspended below the plate, with needle-locking device operated from below. Pivoted at the centre, and freely rotatable, is an alidade with telescopic sights and cross-wires in the focus. The fine scales on the fiducial edges of the alidade are read against the degree scale (0–190) engraved on the raised semi-circular edge of the instrument. A second (fixed) telescope is mounted below the plate, and the whole instrument mounted on a rotatable stage with spring-loaded slow-motion work operated by a detachable knurled knob. This in turn is mounted on a universal ball-joint with a wing-headed screw.

In a fitted leather case lined with red velvet, with tooled *fleur de lys* decoration and brass catches, lock and hinges. The base of the box is fitted with brass studs.

Water level signed 'Lennel à la Sphère à Paris 1774', diameter of tube 18 mm (¾ in.), brass.

Three brass tubes, of which one, mounted on a universal staff-head, provides the centrepiece and has numbered ends to guide attachment of the second and third tubes which have right angle turned ends. To the vertical ends of these tubes holders carrying open-ended glass phials (with brass covers) can be attached. With three corks, in fitted leather case lined with green hessian, tooled *fleurs de lys* decoration, brass catches, hinges and studs.

A pupil of the noted maker Canivet (*fl.* 1750–74) who continued the important workshop of his uncle Claude Langlois (*fl.* 1730–50), Lennel succeeded his master in 1774, the earliest reference to him being in a report presented to the Académie des Sciences in February of that year, which noted his skill, reputation and the fact that he had visited England. He used Canivet's sign, but worked from a different address 'Quai de l'Ecole entre le Café de Parnasse et la Miroitier'. He described himself in 1781 as *Ingenieur du Roi et de la Marine*. In that year he was commissioned to look after the Paris Observatory instruments by Cassini. He had died by 1784.

Fig 193 French (?) geodetic quadrant, not
signed, late eighteenth century, radius
320 mm (12⅝ in.), brass.

Four-strutted quadrant with degree
scale on limb; radial index arm with
clamping and slow-motion tangent
screw, vernier and sighting telescope.

Beneath the plate is a second sighting
telescope for base line observations
and spirit level. The whole instrument
is mounted on an adjustable column
which may be attached to a parallel
plate with four adjustment screws. In
original fitted leather case with *fleur
de lys* decoration.

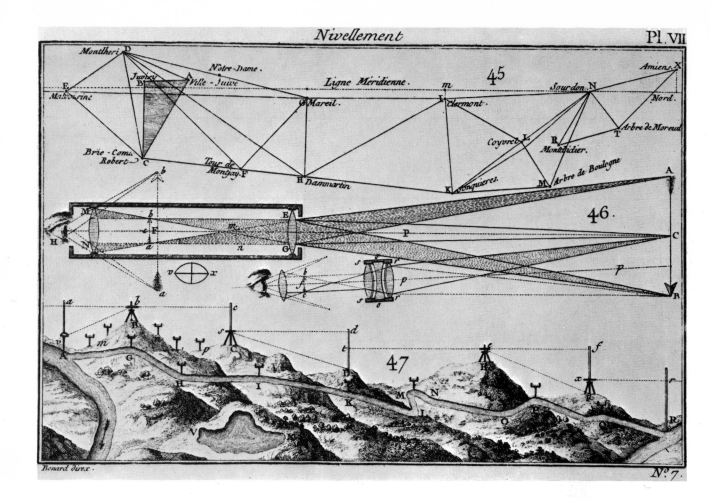

Pl. VII

Fig 194
Levelling, engraving by Benard from
Picard *Traité du nivellement, avec*
une relation raisonée de diverse
nivellements **new edition, Paris, 1780,**
plate vii.

registered. John Love wrote a guide to young colonial surveyors, the *Geodesia* written in 1688 which ran to eleven editions by 1792, with the twelfth and thirteenth editions to follow in the United States in 1793 and 1796.

Vernier scale and telescopes

Two great inventions of the early seventeenth century were the telescope and the vernier, but neither was applied to surveying instruments very quickly. The telescope and the achromatic lens is discussed in chapter 5.

The vernier, an indispensable device on the scale of a sighting instrument, was proposed by Pierre Vernier in *La construction l'usage de mathematique* (Brussels 1631). This device was a scale of divisions attached to an index arm, enabling readings to be taken within narrower limits of accuracy. John Bird (1709–76) improved the vernier by developing a tangent screw, which was a slow-motion precision screw for fine adjustment.

The clamping screw was introduced on the back of the vernier to lock the vernier reading in position while the results could be noted. This was particularly important on a portable instrument used single handed, and on heavy instruments several screws were provided.

The development of the telescope is discussed in detail in chapter 5. Its potential for surveying was quickly grasped and devices were invented to adapt it. An early one was the micrometer, invented in 1638 by William Gascoigne (1612–44), a Royalist who was killed at the battle of Marston Moor. The rationale of the micrometer was that two adjustable metal plates of wires, placed parallel in the focus of the eyepiece of a telescope, could be moved so that the image of the object would be enclosed between them. A precision screw attached to a finely divided scale, regulated these parallel plates or wires so that the distance between them could be read with accuracy, in degrees, thus giving an angular measurement of the size of the object observed.

Fig 195
English sighting level, signed 'Tho. Hogben Fecit, Surveyor, Smarden.', early eighteenth century, length of sighting tube 298 mm (11¾ in.), brass.

Sighting telescope with extended tube at the eye-end, mounted on a horizontal bar, with the spirit level suspended below by two screws. The whole instrument is mounted on a parallel plate with four levelling screws and provision for attachment to a stand or tripod. With original green-velvet-lined fish-skin case.

Fig 197 *below*
English level, signed 'COTTAM & HALLEN WINSLEY ST LONDON', *c.* 1820, length of arm 137 cm (54 in.), overall length of vertical 155 cm (61 in.) mahogany, brass and iron.

Solid vertical member with iron-cased spike and plummet, carrying at its upper end a horizontal sighting arm. Attached to the arm is a semi-circular arc divided with scales 36—0—36 marked 'Fall in inches in 3 feet' and 'inches in 3 feet rise', read against a pointer attached to the vertical staff.

Fig 196
English level, signed 'W & S JONES 30 HOLBORN LONDON', *c.* 1810, length of leg 179 cm (70½ in.), mahogany, boxwood and brass.

Two legs joined at one end, one of them graduated 1—12 in inch divisions followed by 1—4 in twelve-inch divisions towards the joint. At the end of the 1—4 scale is an inset scale 6—0 and 0—6 marked Denton's level labelled on either side of a central ivory slider with sighting hole 'INCHES RISE IN 5 FEET' and 'INCHES FALL IN 5 FEET'. A hole drilled in the opposite leg is fitted with a brass plate with provision for cross threads, with a hinged flap. Mounted inside this second leg is a hinged cross-piece also graduated for inches rise and fall in 5 feet.

Jean Picard in the Paris Observatory used human hair for telescopic sights in astronomical instruments and was probably the first to use the telescope for surveying in about 1667. It was discovered that when parallel pairs of cross hairs were placed at the focal point of the object glass (objective) and their separations compared with the image of a graduated staff placed at the point from which the measurement was to be made, the distance of the staff from the observer was calculable mathematically and without measurement by physical means. This phenomenon enabled the surveyor not only to chart much faster but to reach hitherto inaccessible areas. The micrometer, is in effect, a contrivance for measuring very small distances in the field of view of the telescope.

First invented by Geminiano Montanari (1633–87) in Italy and termed 'stadia', this invention was shelved until it was re-discovered towards the end of the eighteenth century. James Watt employed the device and in 1778 William Green, published in London *The Description & Use of an Improved Refracting Telescope with Scale for Survey*, in which he described how with a given number of threads or intervals in the focus of a telescope, a given distance on a rod could be read, so ascertaining the distance of the rod from the instrument.

Levelling

On the principles of levelling, the surveyor Frederick W. Sims wrote in 1840:

'Levelling is the art of tracing a line at the surface of the earth which shall cut the directions of gravity everywhere at right angles. If the earth were an extended plane, all lines representing the direction of gravity at every point on its surface would be parallel to each other; but, in consequence of its figure being that of a sphere or globe they everywhere converge to a point within the sphere which is equi-distant from all parts of its surface; or, in other words, the direction of gravity invariably tends towards the centre of the earth, and may be considered as represented by a plumb line when hanging freely, and suspended beyond the sphere of attraction of the surrounding objects.'

Levelling – the term used in surveying for finding a true level on undulating terrain – was carried out up to the seventeenth century in a way which had not changed since classical times, with plummet used with graduated square, quadrant or semi-circle (fig. 185) – a method satisfactory for water channels, short distances and range-finding for artillery. A breakthrough came in 1666, when the Frenchman Melchisédech Thevenot described in a tract the bubble/spirit level which, in principle, is still in use to this day. He gave directions for filling a small glass tube about the size of the little finger, sealed at one end, nearly to the top with spirits of wine, leaving a space before the tube was sealed at the other end for a bubble of air as large as the diameter of the tube when it was held upright. When the tube was placed in the horizontal position a mark was cut across the centre of the longitudinal axis of the bubble. The principle was eagerly adopted in England by Robert Hooke who was always looking for new ideas, and he applied it by introducing an air bubble into the centre of a curved plate of glass, in which form it was indispensable for a number of instruments.

The bubble/spirit level was slow to reach the surveyor, as it was difficult to manufacture. Increasing skill in glass-blowing, however, probably stimulated by the immigration of Italian and French workers into England, and familiarity with chemical apparatus and barometer tubes, ultimately made it possible. Whereas in the first four editions of Leybourn's *Compleat Surveyor* there is no mention of development, the 1727 edition describes sophisticated level-readings derived from the bubble level in use since 1679 (fig. 195).

Incline was measured by the clinometer. At first a simple plummet tool, based on the principle of the measurement of the angle between the

horizontal and the incline, by the end of the eighteenth century it had become a sophisticated device, with sights, telescopes, bubble levels and verniers (fig. 202).

The pocket sextant was a useful tool for the nineteenth-century surveyor, which he used for a quick meridian bearing. A handy size for the pocket, they were not intended to have the accuracy of a navigation instrument. Comparatively rare, the miniature instrument by Dollond in fig. 198 in its red morocco case must have been a good servant.

Surveyor's and miner's compasses

A series of sturdy compasses suitable for the surveyor and the mining surveyor were made from the sixteenth century onwards.

The simple compass enclosed in a mahogany case, with either a brass or paper rose (fig. 199), was part of a surveyor's standard equipment. A modification was a compass equipped with one set of adjustable folding sight rules sometimes termed a pelorus (fig. 201), which was used for taking a bearing, and checking other compasses for deviation, as has already been discussed in chapter 2. Miner's compasses fulfilled a vital function in that bearings could be taken both above and below ground to determine the linear direction of mineral deposits and to lay down on paper the position and extent of mine workings.

A peculiarity common to all was that the cardinal points of north and south were also indicated by a ridge on the scale so that they could be told in the dark. On simple types, the glass was traditionally held in place with wax, where indentations with the thumb nail would indicate the bearing required.

Fig 198
English pocket sextant, signed 'Dollond London', c.1790, radius 65 mm (2½ in.), brass, with red morocco case.

A standard form of sextant with silver scale divided 0°–130°. The index arm is adjusted by a tangent screw along the toothed outer edge of the scale arc and has a vernier with a lens mounted in a hinged arm, for reading the scale. With two shades for the horizon glass, hinged wood handle, in original velvet-lined box.

Fig 199
French surveyor's compass, signed
'N. Bion à Paris', *c.* 1700, length of side
of square 95 mm (3¾ in.), brass.

Compass with 8-point rose, and lines
for eight further directions,
surrounded by a raised circumference
engraved with a degree scale
scale (0–90 × 4). Blue-steel needle,
with locking device attached and
controlled by a screw on the plate
outside the box. Compass box with
glazed lid.

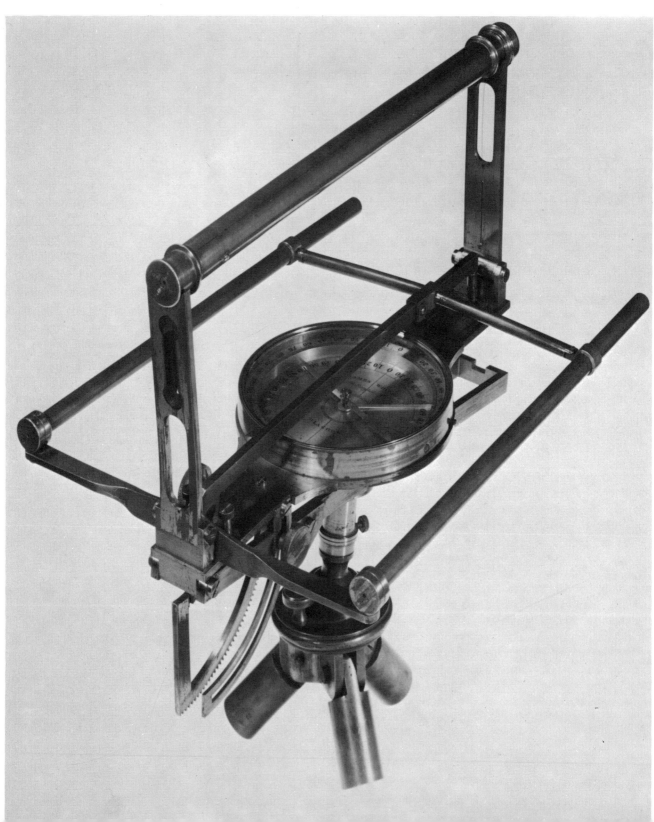

Fig 200 *right*
Scottish surveying compass, signed
'Miller & Adie Edinburgh', *c.* 1820,
brass and silvered brass.

Circular brass box with glazed top and
overlapping base engraved with a
degree scale (0°–360°). Inside the box
is a plain steel compass needle, with
locking device, moving over a scale
of degrees 0°–180°/180°–0° counting
from the N and S points. The four
cardinal points only are marked with
initials, the N point also being
indicated by a *fleur de lys*. On a raised
rim surrounding this scale is a second
degree scale numbered 0°–360° anti-
clockwise. With two folding sights
attached to the outside of the box.

Fig 201 *left*
Scottish mining compass, signed
'JAMES WHITE 14 RENFIELD ST, GLASGOW',
c. 1855, brass.

Magnetic compass with bar edge
needle (with locking device) mounted
in a circular glazed box with degree
scale in four quadrants and a second
degree scale in four quadrants on the
raised circumference of the compass
base. The box is mounted at the centre
of an alidade with open sights, and
with a sighting telescope immediately
above them. A rectangular frame
which is placed above the compass
box carries two additional sighting
telescopes. The whole instrument is
mounted on a cradle attached by a
column and ball joint to a tripod head
with book-plate joints for the legs. The
cradle carrying the compass may be
inclined by a pinion against a curved
rack.

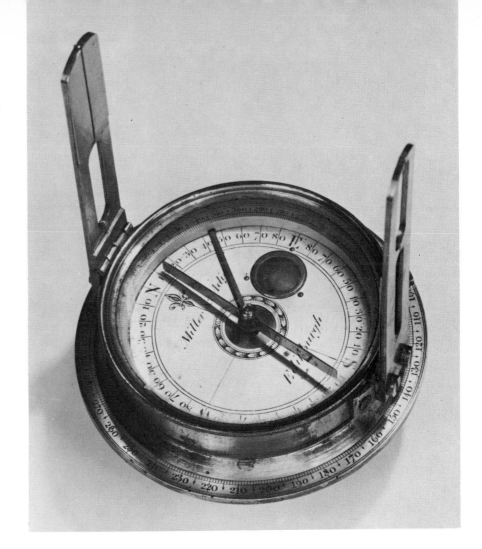

Fig 202 *right*
French clinometer, signed 'Rochette
Jⁿ au Griffon quai de l'horloge à Paris',
early nineteenth century, overall
length 567 mm (22¼ in.), brass, glass
and steel.

Central T-shaped steel arm mounted
on a brass pedestal which incorporates
a spring-mounted screw-thread for
vertical adjustment. Pivoted at one
end, and on either side, of the cross-
piece are two steel arms which carry
vernier scales with fine adjustment.
These move over an unnumbered
scale engraved on a brass limb also
attached to the cross-piece. Brass-
cased spirit-level tubes are attached
centrally to the two arms, which may
be set in approximate positions and
clamped before final adjustment.
Attached to the second arm by U-
shaped brackets is a sighting
telescope.

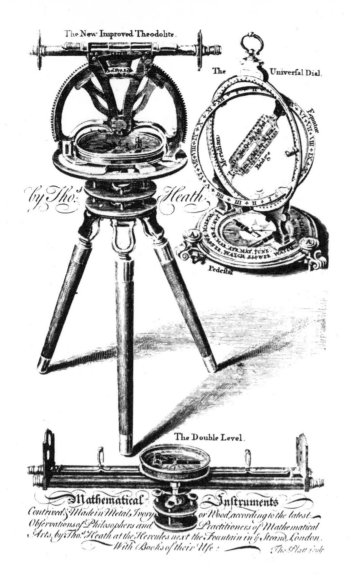

THE
PRACTICAL SURVEYOR:
CONTAINING
The moſt approved Methods
John FOR *Glayter*
Surveying of LANDS and WATERS,
By the ſeveral
INSTRUMENTS NOW IN USE:
Particularly exemplified with
The Common and New Theodolites.
AND ALSO
How to plot and caſt up ſuch Surveys, with the
Manner of adorning the MAPS thereof.
To which are added,
Some uſes of the new Theodolite, *viz.*
In drawing the perſpective Appearance of Beildings, &c.
In levelling, for the conducting of Water, and
In taking the Dimenſions of ſtanding Timber.
Together with the Deſcription and Uſe of
An improved Sliding-Rule for Timber, &c.
An Univerſal Dial.
A Meaſuring Wheel, and
The Pantographer, for copying of Drawings.
Firſt publiſhed in part,
By *JOHN HAMMOND*;
Since enlarged,
By *SAMUEL WARNER*;
And now reviſed, corrected, and greatly augmented.
The THIRD EDITION.
LONDON:
Printed for T. HEATH, Mathematical-Inſtrument-Maker, at the
Hercules and *Globe*, near *Exeter-Exchange* in the *Strand.*
M DCC L.

**Fig 203
Title page and frontispiece from John
Hammonds *The Practical Surveyor*...
third edition, 1750.**

First published in 1731 and possibly in
fact written by Samuel Cunn, this
edition of the book was published by
the instrument-maker Thomas Heath
and describes the surveying
instruments that he sold. The
universal dial whose uses for
surveying are described may be
compared with that shown in fig. 145.

**Fig 204
English theodolite, signed 'Barker',
c. 1780, length of telescope (closed)
330 mm (13 in.), brass.**

Circular base engraved round the
edge with a degree scale, with levelling
screws and spirit level, carrying at
the centre a silvered compass with
8-point rose and needle-locking device.
An inverted V-shaped standard
bridges the compass and carries
between its limbs a toothed and
graduated (70°–0°–70°) semi-circular
arc adjustable by an endless screw.
Mounted firmly to this arc is the
telescope with rack and pinion
focusing and carrying a spirit level.
The whole instrument may be
mounted on a universal staff-head.

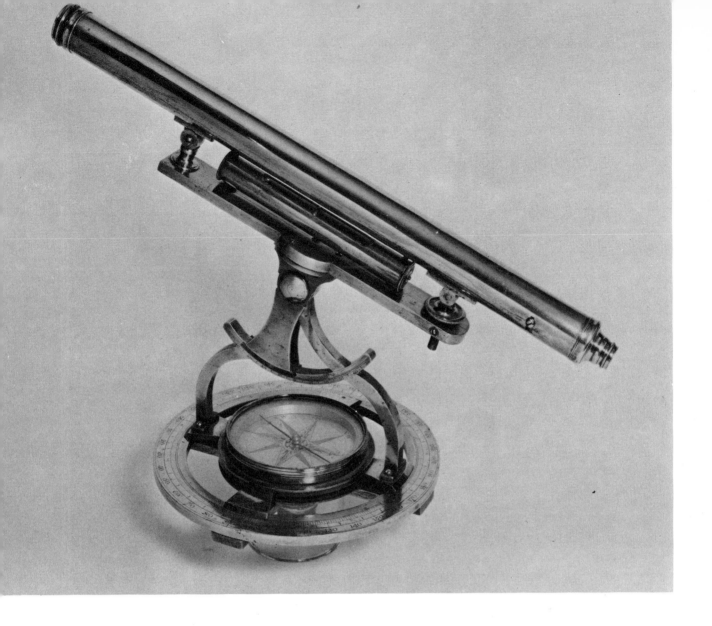

Fig 205
English theodolite, signed 'Hart Fecit
BIRMING^m, *c.* 1800, height from base
plate 245 mm (9⅝ in.), brass.

Circular cut-away base with 360°
scale carrying a central silvered brass
compass in glazed brass box repeating
the signature in Roman capitals and
with named 8-point compass rose.
Mounted on the index arm rotating
over the scale is a bridge carrying the
sighting telescope and elevation/
depression scale graduated 50°–0°–
50°. The telescope has a sliding cover
at each end and is adjusted by a
gnurled nut and screw. A bubble level
is mounted centrally immediately
beneath it.

The outer scale of the compass was divided into 360° and the inner quadrantally, that is to say, in four quadrants of 0–90°. The practical miner's compass which evolved was contained in a mahogany box sometimes fitted with bubble levels with two sights fixed at south and north. One sight vane bears a fine slit in the perpendicular and the other bears a wider slit with a wire extended longitudinally at its centre. The compass box is attached by a universal joint to the tripod, which can be regulated to line up with the horizontal plane (fig. 201).

Later theodolites

In the 1727 edition of Leybourne's book Samuel Cunn complained of the poor quality of theodolites, which had not been improved since those made by Leonard Digges around 1555. However, in the appendix he states, 'I have seen a *Theodolite* made by Mr Sisson, *Mathematical Instrument Maker* at the Corner of *Beaufort Buildings* in the *Strand*, for Accuracy and Dispatch, fitter for a Surveyor than any other I have yet seen.'

Made about 1722, Sisson's theodolite was compact and sturdy. A short telescope with a bubble level attached was fitted to an altitude-adjusting worm-wheeled arc, cut with a clockmaker's culling engine mounted onto a horizontal base with a pinion to set the vernier when the limb was rotated for azimuth. The scales were divided by hand, and read to 6 minutes of arc. John Bird, like Sisson, also made theodolites and was responsible for some of the instruments used by the surveyors Charles Mason and Jeremiah Dixon when they began marking their famous boundary between Maryland and Pennsylvania in 1763.

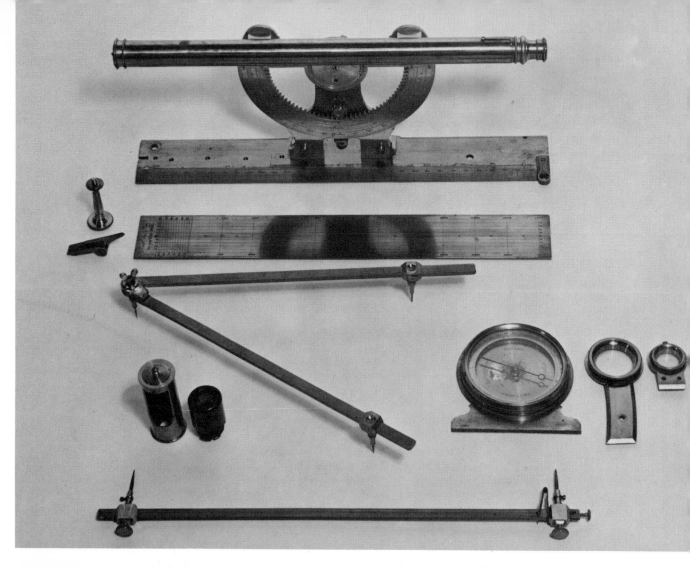

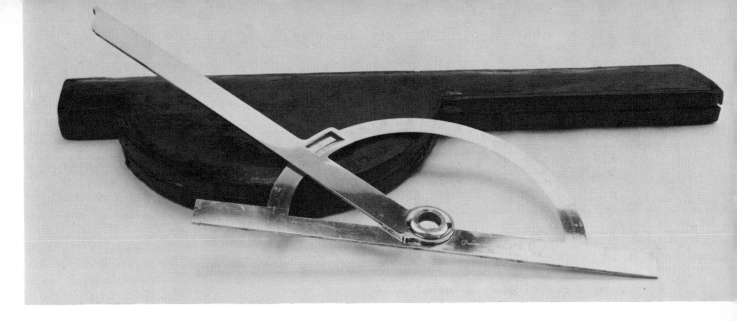

Fig 208
French protractor, signed 'Gourdin au
Quart de Cercle à Paris 1778', length
of arm 402 mm (15⅞ in.), brass in
tooled-leather-covered, fitted wooden
box. Semi-circular arc with double
scale 0°–180°/180°–0° read against a
vernier on the movable arm.

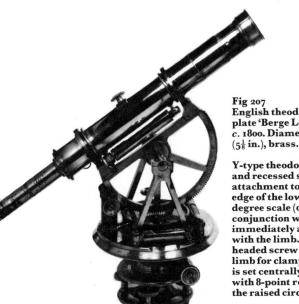

Fig 207
English theodolite, signed on the base
plate 'Berge London late Ramsden',
c. 1800. Diameter of limb 130 mm
(5⅛ in.), brass.

Y-type theodolite with parallel plates,
and recessed screw thread for
attachment to staff. On the bevelled
edge of the lower plate or limb is a
degree scale (0–180 × 2) read in
conjunction with the vernier plate set
immediately above and concentric
with the limb. A collar and wing-
headed screw is supplied below the
limb for clamping. A silvered compass
is set centrally on the vernier-plate
with 8-point rose and degree scale on
the raised circumference. Two spirit

levels are also carried on the vernier-
plate. The toothed semi-circular
vertical arc of the instrument,
adjustable by a pinion, carries on one
side a degree scale 70–0–70 read
against a fixed vernier. On the other
side are two scales 30–0–30 and
100–0–100 marked respectively 'Diff.
of Hypo. & Base' and 'Perp^s in 100^{ths} of
Base' for direct height and distance
measures. The telescope with rack-
and-pinion focusing is carried by two
Y-shaped supports with hinged clips,
each fastened by a pin, itself attached
to the instrument by a cord. Mounted
below the telescope between the
supports is a spirit level enclosed in a
brass canister.

Fig 206 *left*
Set of surveying instruments, signed
'G: F: Brander, Fecit Aug: Vind:' mid-
eighteenth century, length of box
380 mm (15 in.), brass with mahogany
box.

The set includes
Sighting telescope mounted on a
vertical semi-circular ring with triple
degree scale adjusted by a curved rod
and pinion mounted on a base
graduated 0–310.
Proportional rule graduated for
'Pied du Roy'. Beam compass with
fine adjustment. Surveyor's magnetic
compass in glazed brass box. Right
angle lens in fitted original box.

Georg Friedrich Brander (1713–1783)
was born in Regensburg and worked
in Augsburg from 1737.

Theodolites carried telescopes, verniers, adjustable staff heads for
attachment to the tripod, and bubble levels wherever a horizontal plane
was required. Two instruments illustrated in figs. 204 and 205 by pro-
vincial English makers of the fourth quarter of the eighteenth century
embody these refinements.

A greater variety of instruments became available to the surveyor
towards the end of the eighteenth century. The *primum mobile* was the work
of Jesse Ramsden (1735–1800) who produced an historic dividing engine in
1775 which revolutionized scale divisions, previously laboriously executed
with beam compass and dividers. As he had received £615 from the
Government through the Board of Longitude for the invention, he was
unable to take out a patent, and was obliged to publish the details, which
he did in a pamphlet in 1777. Subsequently many engines were constructed in England and on the continent, with modifications, embodying
the basic principles. From these, instruments requiring scales could be
manufactured commercially, and the engine anticipated all the mechani-
zation that was to follow.

Fig 209 *above*
English altitude theodolite, signed
'Abraham Liverpool', *c.* 1800, brass.

Circular plate with extended arms to
carry the open sights. In the centre is
a silvered compass in glazed brass box
with lid on which is engraved a scale
for the difference of hypotenuse and
and base 40–0–40, 30–0–30. 8-point
compass-rose, and needle-locking
device. To one side of the compass is a
spirit level, and surrounding it a
strutted semi-circle engraved on the
arc, which is toothed on its outer edge,
with a double scale of feet 0–6–0 and
6–0–6 marked respectively horizontal
and perpendicular. An index arm is
mounted at the centre of the semi-
circle, with vernier, and carrying the
sighting arm with a bubble level
mounted centrally between the sights
which have cross-wires.

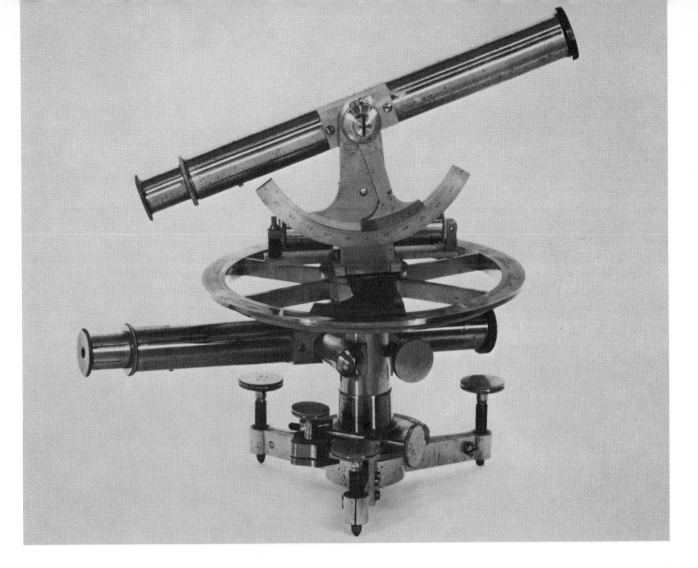

Fig 210
French theodolite, signed 'Lerebours et Secretan, à Paris', overall height 210 mm (8¼ in.), brass.

Fig 211 *left*
English theodolite, signed on the base plate 'Norrien St James Street London' and on the sighting telescope 'Cary London', early nineteenth century, diameter of limb 117 mm (4⅝ in.), brass and steel.

Y-type theodolite mounted on a double plate with three screws. Above this, the lower plate, or limb, has on its bevelled edge a 360° scale for azimuth readings, engraved on steel, made in conjunction with the vernier plate set immediately above, and concentric with the limb, on the vertical axis of the instrument. A silvered compass is carried centrally on the vernier plate, together with a spirit level attached by brackets. The toothed semi-circular vertical arc of the instrument carries a 180° scale (90–0–90) for reading the angle of elevation or depression of the telescope to the mounting of which it is attached. The telescope is carried by two Y-shaped supports with hinged clips, each fastened by a pin. Mounted centrally on the telescope tube, and between the Y-supports, is a spirit-level tube.

In mahogany box with inset blank name-brass.

Ramsden's workshop, headed by his foreman John Berge (1742–1808) who inherited the business when Ramsden died, and who had previously been apprenticed to Peter Dollond with whom he worked until 1791, employed a number of skilful instrument-makers who later became employers in their own right. Of these, François-Antoine Jecker (*fl.* 1790–1820), called 'L'anglais' by his compatriots when he returned to France, turned out the first French mass-produced instruments using Ramsden's dividing engine. He worked with his two brothers, and the instruments are usually not signed. A rare exception, bearing his signature, is an elevation finder, used for measuring the elevation of a distant object in a single operation (fig. 213).

Ramsden's great theodolite was constructed for the Ordnance Survey of Great Britain. In 1784 a common base line for triangulation for both France and Britain was marked on Hounslow Heath. The interest thus aroused, and the example of Ramsden's work, led to improvements in the manufacture of instruments used by land surveyors.

The theodolite signed 'Berge London late Ramsden' in fig. 207 is a typical instrument of the period, and as the nineteenth century progressed, the instruments became smaller with Norrien's (fig. 211) and Adie's. Modifications to the theodolite continued throughout the nineteenth century culminating in machine-made instruments by Stanley, which can still hold their own against modern equipment.

A transitional instrument by Lerebours et Secretan about 1840 embodies some of the old traditions with some of the new (fig. 210).

Surveying of enclosures and counties, followed by international geodetic surveys to align European capitals to meridians, set the old world squarely on the map. The voyages of exploration to the new territories demanded hydrographic skills to survey coastline and chart ocean beds. The instruments used to perform these functions have been described, and by 1800 the growing fund of scientific knowledge was systematized in the great encyclopedias of Chambers, Diderot, Rees and many others.

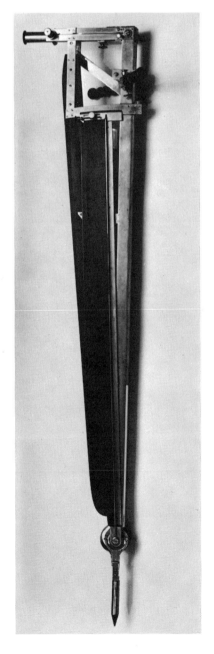

Fig 212 *left*
English alt-azimuth theodolite, signed
'TROUGHTON & SIMMS LONDON' *c.* 1870,
350 mm (13¾ in.), brass and steel.

14-inch alt-azimuth theodolite for
geodetic work which combines
features from the construction of the
transit theodolite with adjustment in
azimuth.

Flat three-footed stand with
adjustable feet, one marked '45'.
Mounted at the centre is a plate
carrying the central column. Rigidly
attached to this column is the base
line telescope. At the top of the
column is a 6-spoke ring carrying a
degree scale 0–360 over which rotates
the index arm with verniers, clamping
and slow-motion screw. A spirit level
is mounted on the index arm together
with a bracket carrying the sighting
telescope adjustable by an index with
fine scale against an altitude arc
60-0-60°.

Pocket globes

Wide public interest in the discoveries being made in astronomy and
geography brought popularity to pocket globes, miniature globes in fish-
skin cases designed to be carried in the pocket. On average 2¾–3 inches in
diameter these spheres could be made of *papier mâché*, wood, or wood with
plaster skins, fitted with mathematically designed gores of printed paper
in which they were completely enveloped. The polar caps consist of circles
of printed paper which cover the edges of the gores. Two pins protrude at
the polar axis of the globes, so that they can revolve freely in the case. The
case consists of two equal cups made of shaped cardboard, stiffened and
hardened with shellac, to which ray skin has been fitted, together with
two brass catches to secure the two halves. Printed gores of the celestial
sphere are fitted inside the cups, to form a vault of the heavens for both
the northern and southern hemispheres. The printed gores are hand-
coloured and varnished.

They would probably be consulted to remind the owner of the geographical location of a port or the identity of a star, or to relate to the
newly published discoveries. There is no evidence to prove that they were
used other than for reference, although a pocket globe at the Museum of
Science in Oxford was at one time pricked out with the points of a pair of
compasses. A peculiarly British invention made in London and seldom
made anywhere else, they were apparently invented by Joseph Moxon
(1627–1700) but exactly when or why is obscure. Born in Wakefield,
Yorkshire, he was in business in 1652 as a globe- and instrument-maker 'at
the Sign of the Atlas' on Parnassus Hill near St Michael's Church in
Cornhill, although in 1670 he was to move his Sign of the Atlas to Great
Russell Street, Bloomsbury and in 1673 to Ludgate Hill near Fleet Bridge.
Moxon turned his attention to printing after setting up business and visited
Amsterdam to study its technology. In 1657 he published an edition of
Edward Wright's *Certain Errors in Navigation Detected and Corrected* and in
1659, based on a publication by W. J. Blaeu, *A Tutor to Astronomy &
Geography, or an easie and speedy way to know the Use of both the Globes, Celestial
& Terrestrial.*

The globes he produced of his own invention were sufficiently important
to rival the publications from the continent so that foreign visitors sought
him out.

Another early maker of pocket globes was Charles Price (*fl.* 1680–1718).
He was apprenticed to a clockmaker, worked as a globe-maker and cartographer, and was associated with several contemporary makers engaged
on similar work. Probably through his connection with John Senex he
made some pocket globes which are now as scarce as those of Moxon. John
Senex (*fl.* 1704–*ob.* 1740; his widow continued with the firm until 1749)

Fig 213
French elevation finder, signed 'Jecker
à Paris', *c.* 1800. Overall height 152 cm
(60 in.), brass and mahogany.

A cylindrical canister containing a
plane-mirror is attached at the lower
end to a reinforced brass limb running
the height of the instrument. This
mirror is fixed to the pivot of a brass
arm, reinforced by a wood strip,
which carries a vernier moving across
a horizontal scale divided 0–12, set at
approximately three quarters of the
height of the instrument. Above the
open top of the mirror-canister is a
hollow brass tube in two portions (the
smaller section being a later repair)
along which the mirror-image is
reflected into a brass box at the top of
the instrument, to a horizontally
placed sighting telescope, with cross-
wires in the eye-piece focus, which
may be adjusted from side to side.

Attached to the telescope box is a
screw-controlled sliding aperture-
cover with triangular slot. Two turned
wood handles are mounted below the
telescope on a collar linking limb and
tube. Screwed to a Y-shaped mount on
the mirror-canister at the bottom is a
brass point for erection of the
instrument.

The function of the instrument would
seem to be to measure the elevation of
a distant object from the observer's
ground level in a single operation.

François-Antoine Jecker (*fl.* 1790–
1820) worked in England with Jesse
Ramsden for five years, before
opening a large *atelier* with his two
brothers in Paris where he utilized
the dividing machines devised by
Ramsden. Although the Jeckers'
workshop was the first in France to
mass produce instruments few signed
examples of their work have survived,
and these, as in the present case, tend
to be the more unusual instruments.

Fig 214 *above*
**English pocket globes, eighteenth
and nineteenth centuries. All of twelve
gores printed from an engraved plate
and coloured by hand, in fish-skin
cases.**

The globes are signed, from left to
right:

Top row Cox, 1835; Cary, 1791; Dudley
Adams, *c.* 1770; Nathaniel Hill, mid-
eighteenth century; Lane (?), 1809.

Bottom row Pair of miniature globes
in turned boxwood cases, signed
'MALTBY'S TERRESTRIAL/CELESTIAL GLOBE
London 1862, E. Stanford, London.'
Overall diameter of boxes, 70 mm
(2¾ in.).

Fig 215 *right*
**English pocket globes, eighteenth
century.**

Top
Signed 'A New & Correct GLOBE of the
Earth by I. Senex F.R.S.' early
eighteenth century, diameter of case
75 mm (3 in.), printed gores coloured
by hand on a central core, fish-skin
case with brass clasps and hinges.

Bottom
Signed 'A Terrestrial GLOBE G: Adams
No. 60 Fleet Street LONDON'.

was a Fellow of the Royal Society and Geographer to Queen Caroline.
He became involved with globe reform along the lines suggested by the
French astronomer Cassini who proposed that they should make compact
globes for wide distribution, rather than the small quantity of magnificent
examples made for royal presentations. In 1712 he worked for a time with
Charles Price and John Maxwell, the three calling themselves 'geo-
graphers'. A globe signed Senex and Price is known and it is believed that
Maxwell also made some. Senex removed his Sign of the Globe to various
premises, including Hemlock Court, near Temple Bar; near the Fleece
Tavern, Cornhill; White Alley, Colman Street; near Salisbury Court,
Fleet Street and ultimately near St Dunstan's Church, Fleet Street. He
was a skilful engraver who made the plates for a number of mathematical
works and also engraved dials and made maps. But his speciality was
globes – of which several have survived and are available to collectors
today.

'A New and Correct Globe of the Earth by I. Senex FRS' shows Anson's
Track and Dampier Straits, Australia connected to New Guinea in the
north, Tasman's Australia marked as 'New Zeland', and some indeter-
minate coastline around the Great Australian Bight. California is shown
as a peninsular of which the northern parts disappear into a void marked
'Incognita'. At a point approximately 52°S there is a circular inscription
'Antipodes of London' (Antipodes – from the Latin *anti* (against) and
Greek *pous* (foot) – those who dwell directly opposite to each other on the
globe so that the soles of their feet occupy diametrically opposite positions).
Japan is vaguely described as two islands with a third entitled 'Bongo' to
the south (fig. 215).

George Adams Sr (*c.* 1704–72) and his sons George (1750–95) and
Dudley (*fl. c.* 1770–*c.* 1810) were the most important of the late-eighteenth-
century instrument-makers. They also sold globes and a pocket globe with
their label appears to have overprinted the plates by Senex. A pocket globe
signed by George Adams corrects Senex's mistake by showing 'New
Holland' (Australia) disconnected from New Guinea, with an entire coast
of New Holland sketched in. There is also a short unconnected coastline
entitled 'New Zealand' separated from the Australian continent, showing
it was made before the three voyages of Captain Cook in 1768–74.
California is connected to Canada and Alaska and the globe lacks the
details of North America later supplied by George Vancouver's *Voyage of
Discovery to the North Pacific Ocean and Around the World . . . 1790–95 in the
'Discovery'*, published in 1798.

Fig 217
English pocket globe, signed 'A New
GLOBE of the Earth by R. Cushee 1731',
diameter of case 75 mm (3 in.), fish-
skin case, gores of paper printed from
engraved plates and coloured by hand.

Fig. 214 shows an unsigned globe with a cartouche proclaiming 'A
Correct Globe with New Discoveries of Dr. Halley & c.' which is attributed
to Dudley Adams. Globes with this slogan were also used by his brother,
George Adams Jnr.

Pocket globes by James Ferguson (1710–76), run to three known
editions. A self-taught youth, he was sent to Edinburgh University by
patrons, and became a prolific writer and teacher of natural philosophy,
able to make his own apparatus and instruments. His first globe, c. 1750,
shows Anson's Track, named after George Anson, who made a voyage in
1740–44 of incredible endurance. Because of the shortage of volunteers,
the Admiralty had ordered the impressment of veterans from Chelsea
Hospital to man his ships; many of these 'volunteers' died before rounding
Cape Horn. Ferguson's pocket globes continued to be published after his
death, probably by Benjamin Martin who bought both Ferguson and
Senex plates.

Richard Cushee (fl. 1708–34) at the sign of the Globe and Sun between
St Dunstan's Church and Chancery Lane, Fleet Street, was associated
with the instrument-maker Thomas Wright from 1731 to 1734. His son
E. Cushee (fl. 1729–68) succeeded him at the Globe and Sun and then
affixed his own sign, The Orrery at Water Lane, Fleet Street. A further
member of the family, Leonard Cushee, worked at the address near
St Dunstan's Church, but is otherwise unrecorded, although globes by
him are known.

'A New Globe of the Earth by R. Cushee 1731' in a scrolled cartouche
shows the west coast of New Holland with Shark's Bay and Lewin
delineated and connected in the north to New Guinea (fig. 217). 'Diemens
Land' is drawn as an incomplete coastline with three small adjacent
islands, and an unconnected coastline between 35° and 40°S is entitled
'N. Zeeland'. California is depicted as an island, and to the north, the
Mississippi River bounds 'Unknown Parts'.

Nathaniel Hill (fl. 1746–66) mathematical instrument-maker, land-
surveyor and globe-maker at the Globe and Sun, Chancery Lane, London,
worked with John Wing on a survey of the Bedford Level in 1749. He
issued pocket globes dated 1754 (fig. 214), and as a variation made a
matching celestial globe which he mounted on turned wooden stands with
crossed stretchers. One of these is among the George III instruments now
in the London Science Museum.

'T. PATTRICK AND CO. 1808' is the inscription on a pocket globe shown in
fig. 221 issued by Thomas Pattrick, globe-maker, optician and instrument-
maker of 29 King Street, Covent Garden. He published two pamphlets
which are now at the National Maritime Museum, Greenwich, *Description
of an Improved Armillary Sphere* in 1802, and *On the Nautical Angle Whereby a
Ship's Departure, Meridional Difference of Latitude, etc., are obtained from
Inspection* in 1803, under the patronage of Admiral Sir William Sidney
Smith. According to Taylor (1192) he also made 'Adams' and 'Senex'
globes, although it is not clear whether he had acquired the original plates.

The pocket globe illustrated bears the outline of New Zeeland (sic) and
a vague New Holland showing a definite coastline of New South Wales
after Captain Cook, but without the additional information provided by
Matthew Flinders R.N. (1774–1814) whose voyages in Australian waters
completed the circumnavigation. Flinder's survey of South-East

Fig 218
Holland circle by Metz, c. 1700.

Fig 219
French circumferentor, c. 1730.

Fig 220
Miners' dial, 1583.

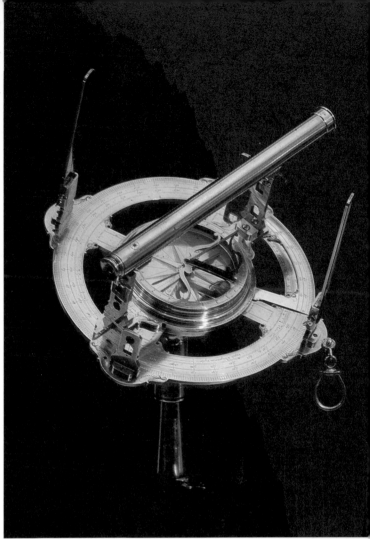

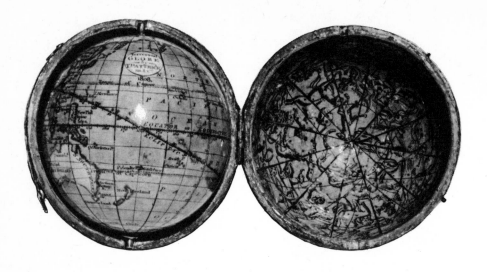

Australia, which is now part of Queensland, and New South Wales in 1796–9 was published in London in 1801 under the title *Chart of Part of the Coast of New South Wales from Ram Head to Northumberland Isles, by M. Flinders, 2nd lieutenant. of H.M.S. RELIANCE*. This double-checked James Cook's survey and was interwoven with corrections. Details may have been used on Pattrick's globe but the scale is too small to be sure.

Flinders made a second world trip in which he circumnavigated Australia in 1802–3, but he and the results of his hydrographic survey were captured by the French off Mauritius on the return journey. While a prisoner between 1804–10 he wrote up his observations, and the French cordially forwarded them to London for him for publication.

Another globe-maker of repute was William Cary (1759–1825), originally an optical instrument-maker who turned to globes. One pocket globe of his dated 1791 has the distinction of being without a celestial map in the case.

Rare pocket globes made by John Addison & Co. (*fl.* 1820–38) of 9 Skinner Street, London (1820); 117 Regent Street, Piccadilly, London (1823); 7 Hampstead Road, Tottenham Court Road (1827) are very occasionally found. Globe-maker to George IV, Addison was one of the principal globe-makers of his time.

One of the leading makers of globes and orreries of the nineteenth century was Newton (*fl.* 1810–1868) of 97 then 66 Chancery Lane, founded by John Newton. The business was later known as Newton & Son, and J. & W. Newton, and by 1838 its title had become Newton, Son & Berry. Without further evidence, one could suppose that the names found on pocket globes of unknown globe-makers may have been printed to order on globes made by Newton.

Of these, Charles Augustus Schmalcalder (*fl.* 1806–38), Optician and Mathematical Instrument Maker at 82 and 399 The Strand, made barometers and optical tools and was the first maker of prismatic compasses. Pocket globes were sold under his name. Pocket globes were issued by N. Lane dated 1776, 1809 and 1818 (the later ones are comparatively frequent) although very little is known about the maker.

No pocket globes seem to have been made later than those of Cox in 1835, which were probably issued (for want of further information) by James Cox, an optician at 5 Barbican, Aldersgate Street, London, between 1830 and 1851. The only known exception is a pair of miniature globes in wood cases entitled 'Malby's Celestial (or Terrestrial) Globe London 1862, E. Stanford, London' made in the style of the small globes for the cabinet produced by Coronelli (fig. 214). Upon close examination the rectangular cartouche has been stuck on over the gores so the question arises as to whether it may conceal another, but most of the modern map has been delineated with certain notable exceptions, so they are of no historical value.

5
Optics

FOR the collector anxious to assemble a collection both comprehensive and representative, optics is a particularly attractive subject. It has developed almost entirely in modern times. The microscope and telescope pre-date the seventeenth century by only a very few years, if at all, and their chief development comes later. The range of optical devices which aid the draughtsman or provide entertainment by reproducing, enlarging or diminishing pictures were nearly all at their most popular in the eighteenth and nineteenth centuries. In consequence they are plentiful, attractive and collectable.

Knowledge of optical phenomena and the geometrical study of vision date far back to the ancient world. Simple lenses made from rock crystal (which occurs naturally) were known by the seventh century B.C. as has been shown by the discovery of a figured lens by Sir Henry Layard during his excavations at Nineveh, and other examples of various dates have been found at Pompeii, Nola and Viking settlements in Götland. The purpose of such lenses is not entirely clear. In Mediterranean regions they may have been used as burning glasses, and in *The Clouds* (*c.* 400 B.C.) Aristophanes alludes to such a use. In more northerly latitudes, however, such functions are less obvious, and while lenses may have found a specialized use in the hands of goldsmiths for magnifying very fine and close work, their more usual role may primarily have been decorative.

Spectacles

The origin of spectacles, like that of the microscope, telescope, compass, and dozens of other common objects, is lost in obscurity. The earliest reliable reference to them is contained in a Venetian guild regulation of 1300. A remark by Friar Giordano da Rivolta in a sermon delivered 23 February 1306 that, 'It is not yet twenty years since there was found the art of making eye glasses which make for good vision . . .' allows the date of their invention to be pushed back to around 1286. It is probable that they appeared first in Italy, perhaps at Pisa, and that they were invented by a lay artisan who was, not unreasonably, reluctant to impart the method of making them to anyone else. After 1300 reliable references to eye-glasses are sufficiently numerous to show that they quickly became a valuable adjunct for the literate classes as with old age their sight failed and reading became difficult. Towards the end of his life, the poet Petrarch (d. 1374) noted that he had to seek the help of eye-glasses. These were used not as a corrective for faults in vision, but simply to enlarge and so to allow the long-sighted or the weak-eyed to read more easily.

The form of early glasses was very simple. The two lenses were mounted in a rim of horn or leather with a short straight projecting arm. These arms were riveted together allowing the lenses to be folded over each other. There were no side pieces since the glasses were not intended to be worn continuously but merely for reading or other close work.

Even in medieval times when there was a certain amount of variation in the form of the frames, there were no fundamental changes. Lenses might be made either of beryl, quartz (rock crystal) or glass, but they were always convex and only assisted the long-sighted. Not until the end of the fifteenth and the beginning of the sixteenth centuries were concave lenses introduced to help the myopic, and it was not until the astronomer Johannes Kepler had worked on the refraction of light by the eye that the beginnings of a scientific understanding of how spectacles worked were laid.

The sixteenth century did, however, see considerable variety in both the forms and the materials of which eye-glass frames were made. Also, through mass-production, it witnessed a reduction in price. Horn, bone, leather, ivory, gold, silver or iron could all be used for the frames, which were often elaborately decorated. Cords and leather straps, and single clips from the bridge of the spectacles to the wearer's hat, were used to hold the glasses in place. This problem was partly lessened by the introduction, early in the seventeenth century, of the spring bridge, but it was not for

Fig 224
Portrait of John Cuff and an assistant.
Signed 'Zoffany pinx/1772', 895 × 692 mm (35¼ × 27¼ in.), oil on canvas.

A precisely detailed portrait of John Cuff at his work bench. In his hand he holds a piece of crystal while immediately beside him is a small polishing machine. Exhibited at the Royal Academy in 1772 (no. 291), Horace Walpole pencilled into his catalogue the comment 'extremely natural but the characters too common nature, and the chiaroscuro destroyed by his servility in imitating the reflections of the glasses'.

Fig 222 *(previous page)*
Portrait of an unnamed gentleman with telescope, spectacles and quizzing glass, c. 1800.

Fig 223 *(previous page)*
Reflecting telescope signed 'Paris à Paris', late eighteenth century.

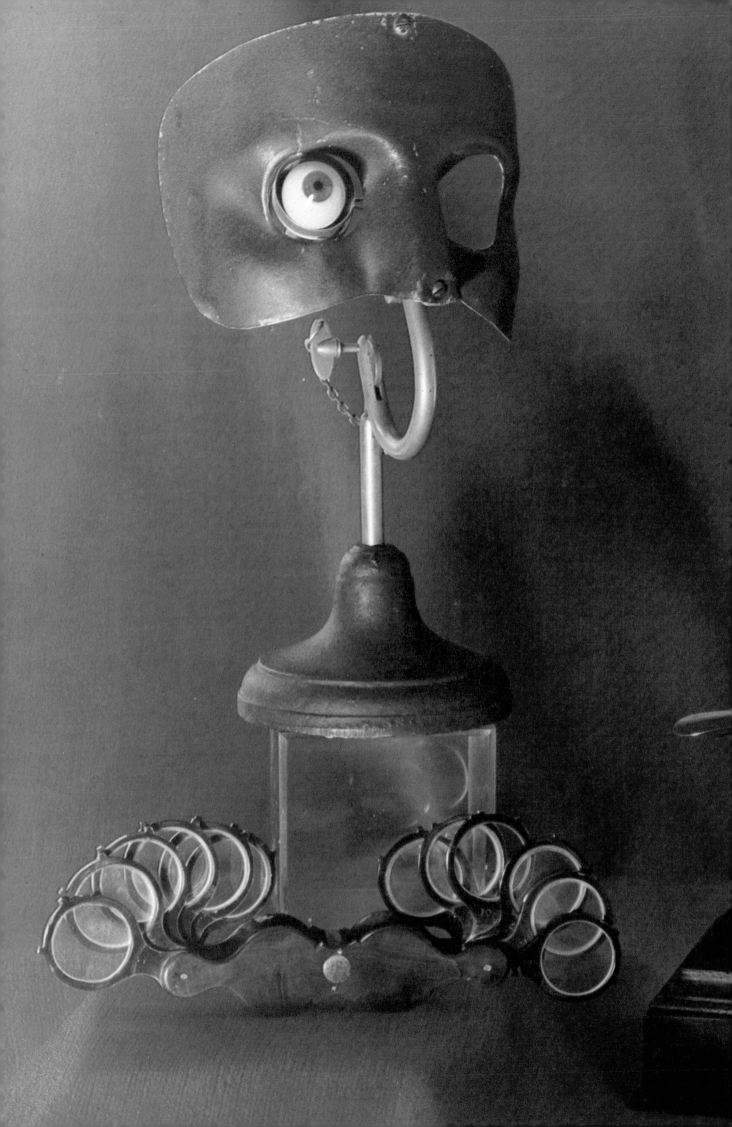

another century that the more practical solution of attaching rigid side pieces to the lens-holders was introduced, probably in London. The optician Edward Scarlett (*c.* 1688–1743) is often associated with the popularization of eye-glasses of this kind, which were commonly known as temple-spectacles because the side pieces ended in rings which pressed against the temples, rather than clipping over the ears like modern spectacles. Soon after, spectacles were introduced with double-hinged side pieces which could return behind a wearer's wig. 'Wig-spectacles' of this kind were advertised by James Ayscough in 1752.

After the invention of side pieces, frames developed considerably, and the inventive London opticians of the eighteenth century, including many of the best-known makers of optical instruments such as Benjamin Martin and Dudley Adams, experimented with new forms. Examples of a form devised by Adams, in which the lenses hung in front of the eyes from a rod set across the forehead, may be seen in the Science Museum, London. Attempts were also made to overcome the irritation caused by the need to change spectacles when reading in order to look into the distance. Benjamin Franklin (1706–90) is generally credited with the invention of bi-focals, which helped to overcome this problem, although the evidence available does not put his authorship beyond doubt. Certainly Franklin's letters supply the earliest descriptions of bi-focals, but it is significant that his phrases imply that they existed earlier. Moreover, Sir Joshua Reynolds may already have been wearing bi-focals in the 1780s, as was Benjamin West before the end of the eighteenth century.

Possible alternatives to bi-focals which are sometimes found were double glasses in which an additional pair of rimmed lenses was hinged to the frame either from above, as in the early forms patented by Addison Smith (1783) and J. R. Richardson (1797), or from the side. Glasses of this kind, which enjoyed some popularity in the early nineteenth century, often used the second rim to hold a tinted glass rather than a magnifying lens. The shape of lenses also exhibited more variety in the nineteenth century, oval and rectangular forms becoming popular with silver or tortoiseshell often used for the frames. Cheaper pairs of spectacles were made of steel, and rimless spectacles which appeared around 1830 became popular from the middle of the century. Characteristic of the second half of the century was the pince-nez.

Fig 226
Two pairs of spectacles, early nineteenth century.

Left
Silver, with initials 'T P', and hinged side-pieces. Silver-mounted shagreen case with crest of a cockerel bestriding a coronet, length 115 mm (4½ in.).

Right
Gold-wire frames with ear-pieces; in silver-mounted, tortoiseshell case, length 110 mm (4¼ in.).

Fig 227 *(overleaf)*
Small telescopes, nineteenth century.

Fig 228 *(overleaf)*
Spy glass, nineteenth century.

Scientific understanding of the action of spectacles developed slowly. Astigmatism was among the earliest defects of the eye to be studied exhaustively, the work being carried out by Sir George Airy, Astronomer Royal, in around 1827. Further investigation of the eye fell later in the century and is beyond the scope of this book. Of some interest, however, are the various instruments and tools developed to aid the optician either in the manufacture or the fitting of spectacles. Lens-grinding lathes, specially divided rulers, spherometers for determining the refractive value of a lens from the curvature of its surface, and the apodemeter for determining the strength of glasses needed by a customer are among the devices which may occasionally be found. Few of these are likely to date from earlier than the middle of the nineteenth century.

The Telescope

'That Galileo did not invent the refracting telescope (in the usually accepted sense of the word) is largely irrelevant in an assessment of the part played by the telescope in scientific investigation and its effect on thought and attitudes' (F. R. Maddison *Galileo*). Beyond that its name was invented by the Greek poet and theologian John Demisiani of Cephalonia and first publicly used at a banquet in honour of Galileo on 14 April 1611 by Federico Cesi, president of the Accademia dei Lincei, little is known of the immediate origins of what was to be a revolutionary instrument, the refracting telescope. While several antecedent devices which may have influenced its invention can be found, and examples of the use of lenses demonstrated earlier, no evidence exists to show precisely how it was first produced. Perhaps the most significant fact available is that the men first associated with it, Zacharias Janzoons and Hans Lippershey, were both spectacle-makers and that both sought reward and recompense from the Dutch Estates General on the grounds of its *military* value. So too did Adriaan Anthoniszoon (Metius).

The earliest reference to the telescope, by Isaac Beeckman in his journal in 1634, while mentioning Janzoons's work in 1604, also refers obliquely to a still earlier instrument supposed to have derived from Italy in 1590 – a complication which merely underlines our ignorance. However, what is important is to recognize the non-scientific origin of the telescope, and the rapidity with which news of it spread across Europe. Telescopes were on sale at the Frankfurt fair in autumn 1608, and two of Lippershey's instruments had been sent to Henri IV of France by December 1608. The following spring telescopes were on sale in Paris, but it was not until June 1609 that Galileo heard of them. Soon after he began to attempt to construct one for himself, making it in early July 1609.

The history of the telescope as a scientific instrument begins with Galileo and continues with the work of Thomas Harriot. None of the latter's instruments, made for him by Christopher Tooke, survive, but two telescopes and a broken objective glass ascribed to Galileo are now in the Museum of the History of Science in Florence. These early telescopes consist of wooden tubes covered in paper or leather, and lenses held in position by a wire clip which also holds pasteboard stops in place against them. However, the first telescopes sold in France are known to have had tubes of tin-plate, and Galileo himself referred to a 'Tubum plumbeum'. But use of metals for this purpose seems to have disappeared quickly, wood and/or pasteboard tubes covered with leather and/or paper becoming normal. Tapering instruments with the objective set in the smaller end, or in the smallest draw tube, the reverse of modern practice were quite common. A further form, especially useful for larger telescopes which lacked rigidity, was provided by the use of square section or octagonal tubes of oak or, later, mahogany.

The Galilean (or 'terrestrial') telescope, employed only two lenses: a bi-convex object lens and a bi-concave eye lens. It produced an upright and reasonably bright image in a well-illuminated field (fig. 230). It was therefore particularly suitable for the military and naval uses which were its chief value to contemporary society. A disadvantage was that only a

Fig 229 *(previous page)*
Microscopes, Culpeper, Cuff and Cary
types.

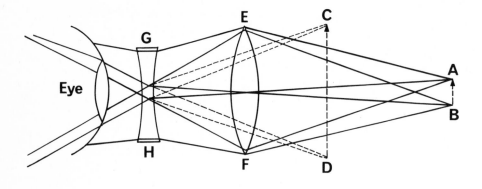

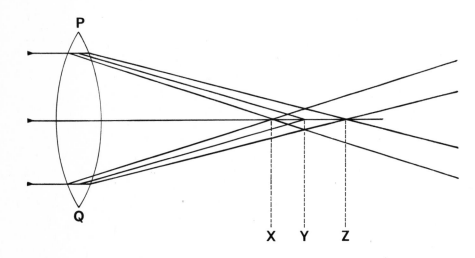

Fig 230
Diagram of the action of the Galilean
telescope.

The convex lens EF makes the rays
from the object AB converge, while
the concave lens GH diverges them
once more and renders them parallel
for reception in the eye. The rays
received in the eye after passing
through GH look as if they are
proceeding along the dotted lines from
an object CD closer to the eye which
thus appears larger.

Fig 231
Diagram to illustrate chromatic
dispersion.

PQ, a bi-convex lens, receives parallel
rays of white light which are refracted
to different foci for their constituent
colours: X for blue rays, Y for yellow
rays, Z for red rays.

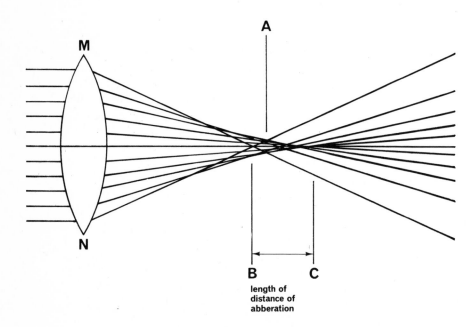

Fig 232
Spherical aberration.

MN convex lens
A area of least confusion
B focus of marginal rays
C focus of axial rays

Because all the rays from an object do
not come to a single focus, the image
produced is poorly defined and cannot
be brought to a sharp focus.

very small field of vision could be seen even when a low magnifying power was used. Furthermore, it suffered from defects caused by chromatic dispersion and spherical aberration (figs. 231 and 232). At the beginning of the seventeenth century the theory of neither of these was understood, but in his *Dioptrice* (1611) Kepler described the action of the eye, and with it the defect of spherical aberration. He suggested how it might be overcome by imparting a hyperbolical curve to the surface of a lens, but the practical difficulties of lens-grinding meant that little use could be made of this advance in understanding. Kepler further suggested in the *Dioptrice* that the field of view of a telescope might be enlarged by using two bi-convex lenses (fig. 234) instead of a bi-concave and a bi-convex, but this had the disadvantage for everyday use that the image then appeared inverted. However, this hardly mattered for astronomers, and the Jesuit astronomer Christoph Scheiner stated in 1630 that he had used such a system in 1619. The arrangement, in fact, came to be known as the 'astronomical' or 'celestial' telescope. Anton Maria Schyrle von Rheita, who discussed the eyepiece in detail in 1645, incidentally forgetting Kepler's part in the invention, adapted it for terrestrial use by adding an erector lens as well as a field lens. With this arrangement a wider field and and an erect image were obtained.

The development of the astronomical telescope in the seventeenth century led to the making of ever-larger instruments as it was discovered that in practice the effects of both spherical and chromatic aberration were reduced as the ratio between the lens aperture and the focal length increased. The unwieldy monsters produced as a result have long since perished, and their history lies outside the scope of this book, as does that of the aerial telescope devised by Huygens as a means of using a long focal length without having to manipulate a heavy, shaking and cumbersome tube. Despite their difficulties it seemed to Newton that there was no 'other means of improving Telescopes by Refractions alone, than that of increasing their lengths' (*Optics*, 1704), and it was this belief which led him directly to investigate the magnifying properties of mirrors. 'Seeing therefore the Improvement of Telescopes of given lengths by Refractions is desperate; I contrived heretofore a Perspective by Reflexion, using instead of an object glass a concave Metal.'

The story of the development of the reflecting telescope is long and complex. It dates back at least to the mid-sixteenth century, and more indirectly into the Middle Ages, but it need not concern us here. The seventeenth century had seen several suggestions for a reflecting telescope, most notably that of the Scotsman James Gregory (1638–75) who published a design in his *Optica Promota* (1663). While in London supervising the publication of this work, he sought to have an instrument of his design made, but abandoned the attempt on failing to find an optician capable of figuring the mirrors to the correct curvature.

Gregory's design for a telescope offered several advantages. By using a primary mirror of concave paraboloid curvature and a concave ellipsoid secondary mirror, he obtained a small, easily managed instrument of high magnification and producing a distinct, clear image (fig. 235). A further advantage, of which Gregory himself was unaware, was that metal mirrors do not suffer from chromatic aberration. This fact did, however, become evident to Newton, who concluded, erroneously, from his experiments on light that refraction of white light was always accompanied by its dispersion into the component colours. This belief, combined with the difficulties of manipulating long-focus telescopes, led him in 1668 to complete a reflecting telescope (figs. 236 and 238). This first instrument was only 6 inches long (153 mm), but in quality it compared favourably with a 4 foot refractor (122 cm). In 1671, Newton's second model was shown at a meeting of the Royal Society in London. The following year news was received from France of a further form of reflecting telescope, invented by a Monsieur Cassegrain (fig. 237) for which priority was claimed. After a devastating reply by Newton this claim was dropped, but the dispute had little importance. Throughout the following century all three kinds of reflecting telescope were made and used.

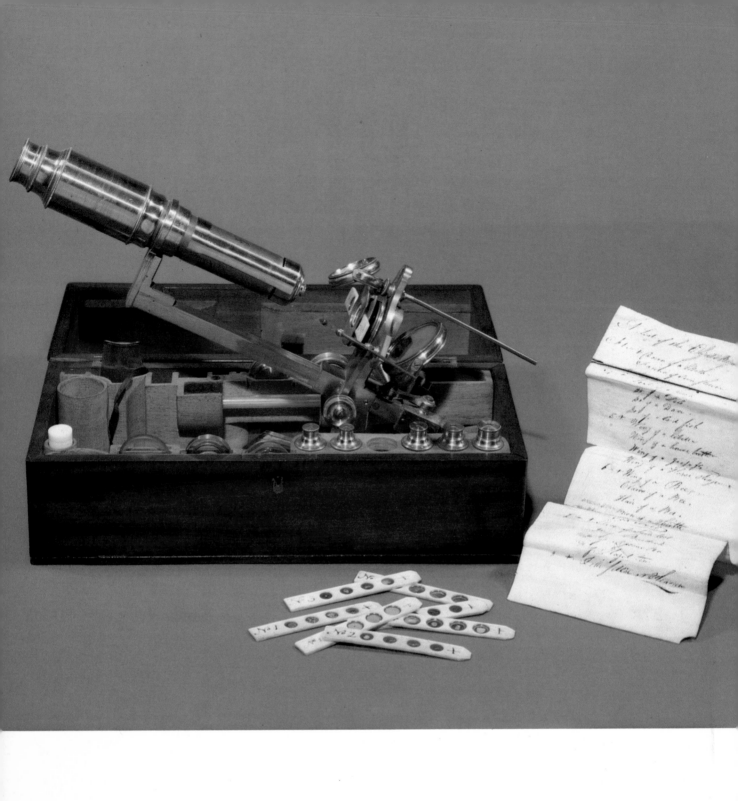

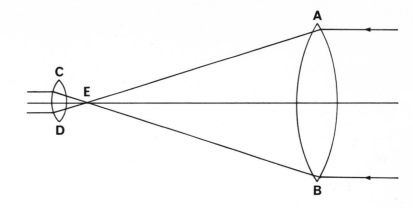

Fig 233 *left*
Chest microscope by Edward Nairne,
c. 1780.

Fig 234
Because the rays refracted through
the objective lens AB cross before
reaching the eye glass CD the image is
seen as inverted.

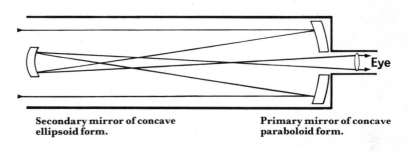

Fig 235
The Gregorian reflecting telescope.

Secondary mirror of concave
ellipsoid form.

Primary mirror of concave
paraboloid form.

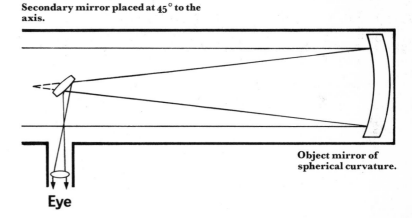

Secondary mirror placed at 45° to the
axis.

Fig 236
The Newtonian reflecting telescope.

Object mirror of
spherical curvature.

Eye

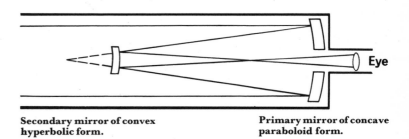

Fig 237
The Cassegrain reflecting telescope.

Secondary mirror of convex
hyperbolic form.

Primary mirror of concave
paraboloid form.

Fig 238
English reflecting telescope, not
signed but by Sir Isaac Newton, 1671,
brass.

The original reflecting telescope made
by Newton with his own hands to
demonstrate his new method of
construction. The instrument was
presented to the Royal Society in 1766
by the instrument makers Heath and
Wing.

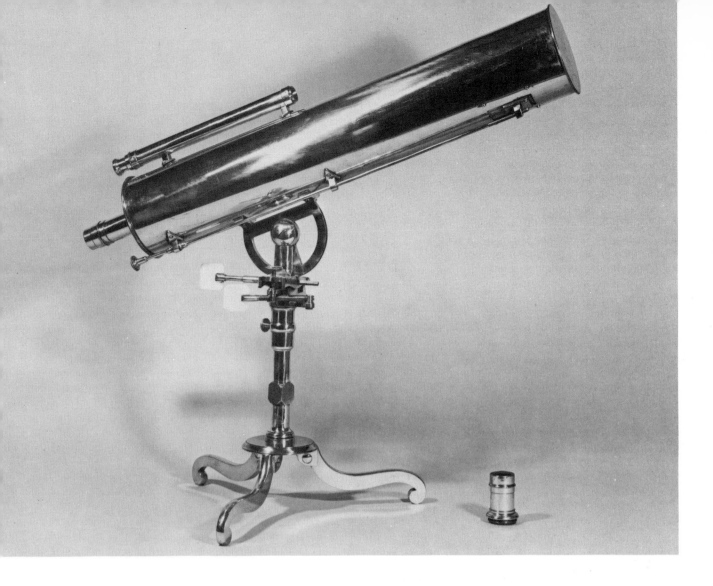

Fig 239
English reflecting telescope, signed
'JAMES SHORT LONDON $\frac{109}{780} = 18$', $c.$ 1754,
length of barrel 603 mm ($23\frac{3}{4}$ in.),
brass.

Eighteen inch focus Gregorian
telescope with geared alt-azimuth
mounting on a pillar and tripod stand.
Fixed focus refracting telescope for
sighting mounted on top of the barrel.
With two eyepieces and two secondary
specula.

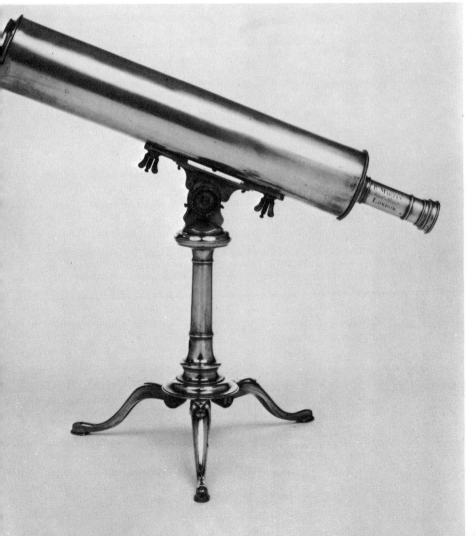

Fig 240
English reflecting telescope, signed
'B. MARTIN, Fleet Street, LONDON', mid-
eighteenth century, length of barrel
456 mm (18 in.), brass.

$2\frac{1}{2}$ in. Gregorian telescope rotating on
a single stem stand with three folding
cabriole-type legs. Detachable
compound eyepiece; screw thread
focusing by a rod attached to the
barrel (not visible in illustration).

195

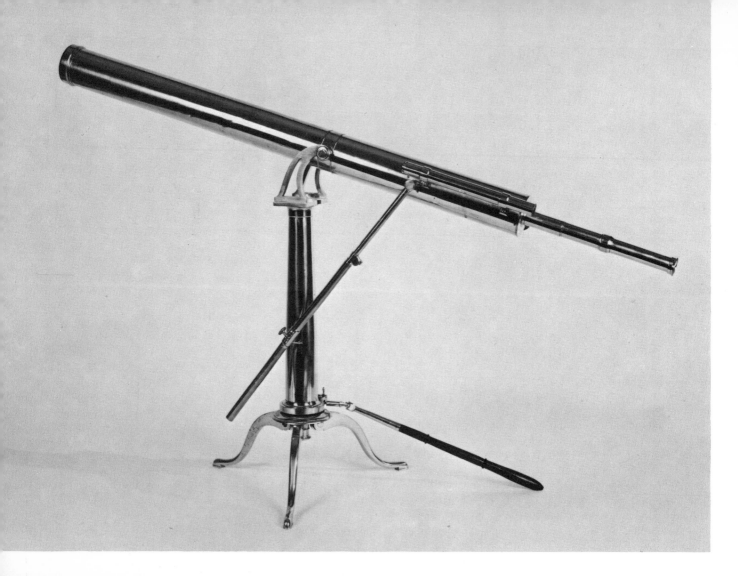

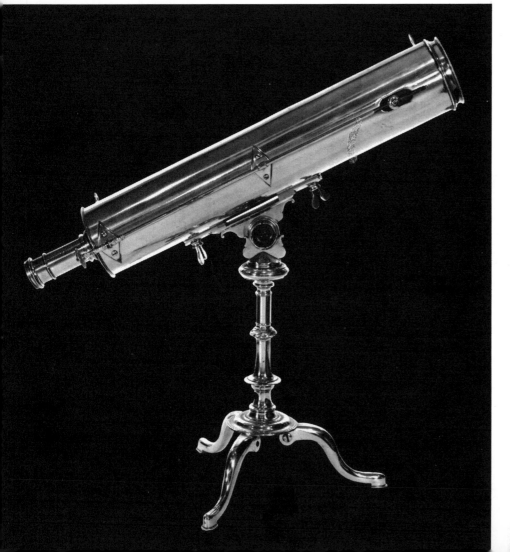

Fig 241
English refracting telescope, signed
'Ramsden London', length of barrel
105 cm (41¼ in.), late eighteenth
century, brass.

3-inch achromatic celestial and
terrestrial telescope swivel mounted
on a tapering column resting on three
cabriole-type legs. With a steadying
rod, and mahogany arm for rotating
in azimuth, articulated by a Hooke-
joint; rack and pinion focusing;
sighting telescope.

Fig 242
English reflecting telescope, signed
'F. HOWEL LONDON', c. 1740, diameter
of objective 76 mm (3 in.), brass.

3-inch alt-azimuth Gregorian
telescope with detachable eyepiece,
open sights and fine focusing by
threaded rod attached to the exterior
of the tube, mounted on an elaborately
turned and unusually slender single
stem stand with folding, tripod feet.

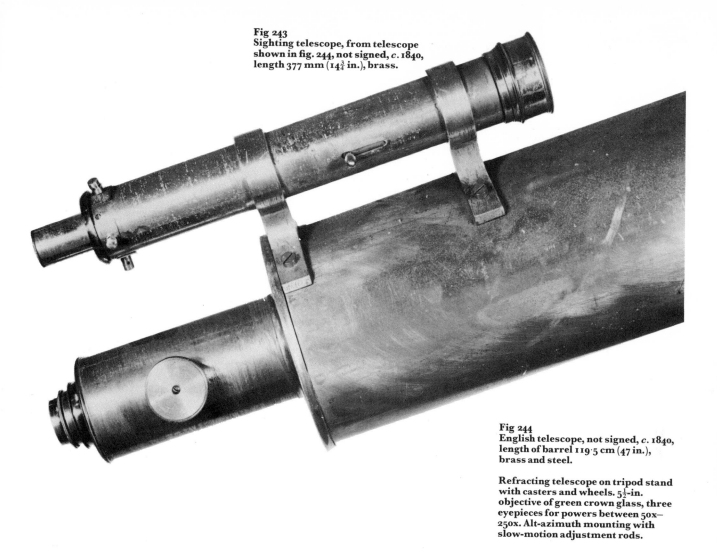

Fig 243
Sighting telescope, from telescope
shown in fig. 244, not signed, c. 1840,
length 377 mm (14¾ in.), brass.

Fig 244
English telescope, not signed, c. 1840,
length of barrel 119·5 cm (47 in.),
brass and steel.

Refracting telescope on tripod stand
with casters and wheels. 5½-in.
objective of green crown glass, three
eyepieces for powers between 50x–
250x. Alt-azimuth mounting with
slow-motion adjustment rods.

Although the advantages of the reflecting telescope were clear after Newton's work, the practical difficulties of making good mirrors were considerable. It was not until the second and third decades of the eighteenth century with the work of John Hadley and Samuel Molyneux that reflecting telescopes started coming into general use. Investigations into the composition of speculum metal (an alloy of copper and tin) led to the establishment of the most satisfactory proportions for the grinding and polishing of mirrors, while casting techniques were also improved, notably by George Hearne. It was he, who with Edward Scarlett, established the techniques of speculum metal mirror-making in the London optical trade, while the publication in 1738 of Robert Smith's *A Compleat System of Opticks* (fig. 248) made a comprehensive account of optical work generally available. Soon other London opticians were making reflectors, and a trade was well established before the Scotsman James Short made his first experiments in 1732.

James Short (1710–68), 'Optician solely for Reflecting Telescopes', was one of the two outstanding telescope-makers who worked in England during the eighteenth century. In a working life of approximately thirty-five years Short produced 1,370 telescopes, mainly Gregorians (fig. 239). Most of them (over 93 per cent) were of small focal length, between 3 inches and 18 inches. It seems unlikely that Short made any part of his telescopes except the mirrors, yet his skill and reputation were such that he was able to charge up to twice as much for instruments as his rivals. He was unusual among his fellow instrument-makers for specializing in one device only, and in not maintaining a very large workshop. His excellent telescopes provided a standard of comparison for his contemporaries, and it is fortunate that many of them have survived.

James Short concentrated upon the most essential part of the telescope, its optical fittings. Almost equally important, however, was the rigidity of the stand, and the manoeuvrability of the mounting. For the modern collector the development of telescope stands, and their variety, offers a very suitable theme on which to centre a collection and one from which much of value may be learnt. For smaller instruments of short focal length, a separate stand might not be necessary, a screw on the end of the pillar which could be placed in a thread set into the centre of the box-lid producing a sufficient mounting. Larger instruments of from 12 inches to 24 inches focal length needed a more elaborate stand, and such instruments were commonly fitted on a heavy brass column carried on a tripod with cabriole-type feet (fig. 240). Mountings themselves, sometimes with fine-screw adjustment, manifest great variety, whether they provided simply for alt-azimuth movement, or movement in the plane of the equator as well (figs. 242 and 241). Larger telescopes were also usually fitted with open (fig. 242) or telescopic (fig. 243) sights for locating the object to be observed.

Just as James Short's telescopes set the standard for his generation, so those of the second great maker of the eighteenth century, William Herschel (1738–1822), set the standards for his. Where Short was a professional optician, whose main activity was in telescope-making, Herschel was a musician whose enthusiasm for natural philosophy in general and astronomy in particular led him to experiment with making telescopes for his own use. The experience he gained in so doing led him to effect astonishing improvements in magnifying power and took him almost accidentally into making telescopes for sale. Following his discovery of the new planet Uranus in 1781 his fame and the remarkable qualities of his mirrors led to a heavy demand for his telescopes despite their high price – £200 for the standard 7-foot model. Comparatively few of them were bought by private individuals and today they are rare outside scientific institutions. But this was not the case with another development made by a non-professional instrument-maker, the invention of the achromatic lens by Chester Moor Hall.

In 1729, Hall, a barrister who carried out optical work for amusement, combined a concave lens of flint glass with a convex lens of crown glass.

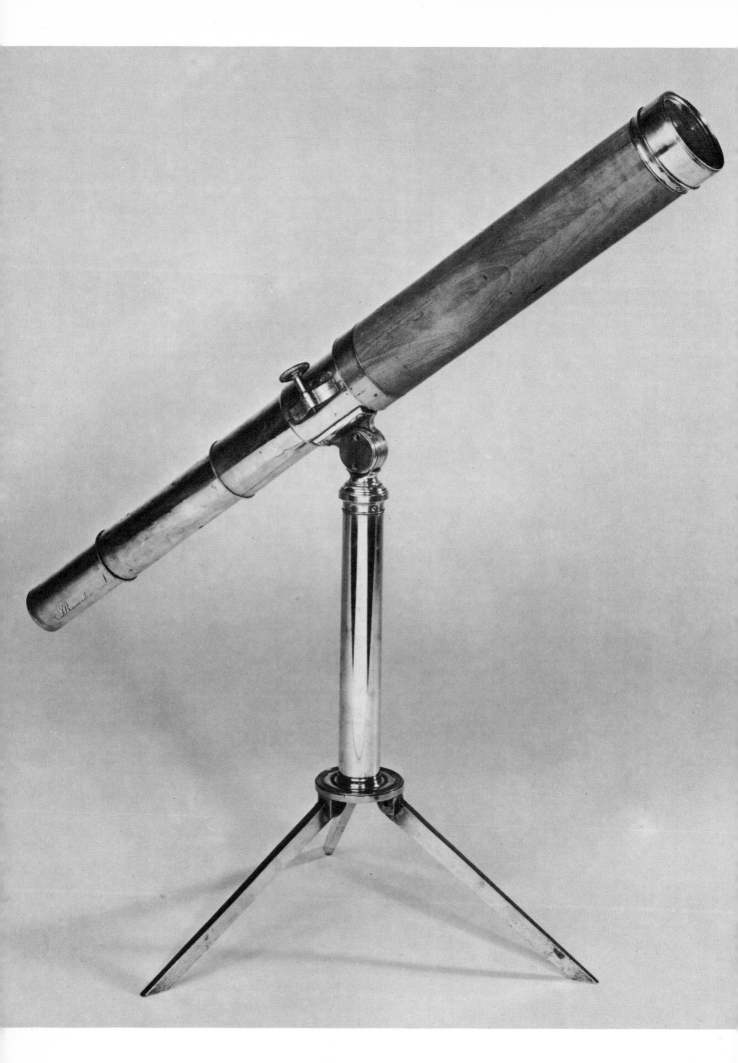

Fig 246 *left*
English refracting telescope, signed
'Ramsden London', eighteenth
century, mahogany and brass.

Four-draw telescope of the hand-held
type which may be mounted on a
straight-leg tripod stand with plain
straight pillar, by an annulus and
screw, which encircles the body.

Fig 247 *right*
Mounting of a 2-inch telescope, signed
'Dollond London', length of telescope
closed 308 mm (12⅛ in.).

A hinged brass annulus with locking
screw is mounted on a simple alt-
azimuth mounting with straight legs.
This annulus may be placed around
the body of a hand-held telescope, thus
converting it to a table model.
Mahogany tube, with four brass draw-
tubes. In fitted mahogany box with
inset brass name-plate (initials
'H S R'). On the underside of the lid is
a trade label of 'William Harris & Son
Manufacturers of Optical,
Mathematical & Philosophical
Instruments, 50 High Holborn corner
of Brownlow Street London' and the
handwritten note 'cleaned & adjusted
Aug^st 1846'.

Fig 248 *below*
English magnifying glass, not signed,
late eighteenth century, diameter
216 mm (8½ in.), mahogany, with
turned screw-in handle.

The resulting compound lens produced a far clearer image than others available, and instructions for making it were given by Hall to various London instrument-makers including John Bird and James Ayscough. His lens was achromatic, but this was not established until John Dollond (1706–1761), a weaver, investigated the matter in far greater detail. In combination with his son Peter (1730–1820), Dollond began to manufacture the new lens under the protection of a patent. As its full importance was recognized, so the patent came under pressure and in 1764 thirty-five London opticians petitioned that it should be revoked. Although the petition failed, it was constantly infringed, and the succeeding years until its expiry were fraught with litigation. Even so, the publication of Dollond's full investigation encouraged much further work in Europe and by 1763 achromatic telescopes were being made in France, Louis xv received one that year from the maker Passement.

The performance of the achromatic telescope was fully comparable with that of the reflector with the result that the latter gradually dropped out of use in the first half of the nineteenth century. The Napoleonic wars, and the initiative of the house of Dollond, also saw a great rise in demand for hand-help telescopes (fig. 249).

Throughout the seventeenth and eighteenth centuries, spy-glasses and telescopes for terrestrial use with from one to eight (or occasionally more) draw-tubes were produced. These instruments were made with pasteboard tubes covered in paper, vellum or leather (fig. 250). A recent study of the decorative tooling on them has shown that while the punches used do not help to identify the makers of unsigned instruments, they do offer some broad guidelines to dating. Although examples of vellum and paper instruments persisted into the nineteenth century, they were increasingly displaced by telescopes with brass-bound mahogany bodies and brass draw-tubes, a form introduced by Dollond in about 1780, or by all-brass instruments. Telescopes of this kind, which could vary considerably in size from just over 1 foot to something over 4 feet extended length, were also supplied with small stands to which the tube could be attached by an annulus (figs. 246 and 247). Some of the larger instruments of this type were also supplied with a celestial eyepiece, while a more rigid form was produced by using an octagonal section tube. Only towards the end of the eighteenth century did the leather-bound brass telescope, which is perhaps that most familiar today, appear in the instrument-makers' lists.

Fig 250 *Above right*
Italian telescopes
Horizontal
Signed 'LEONARDO SEMITECOLO', eighteenth century, overall length closed 410 mm (16¼ in.), pasteboard tubes, tooled outer tube, ebony and bone mounts.

Vertical
Signed 'DA SEMITECOLO VENETZIA', eighteenth century, overall length closed 314 mm (12⅜ in.), three pasteboard tubes, bone mounts.

Fig 249
English hand-held refracting telescopes.

Left Signed 'Dollond', c. 1800, overall length closed, 160 mm (6¼ in.), vellum and rayskin with brass mountings.

Two draw, ½in. telescope.

Right Signed 'Dollond & Son', late eighteenth century, overall length closed 250 mm (9⅞ in.), vellum and rayskin with silver mountings.
Five draw, 1½ in. telescope.

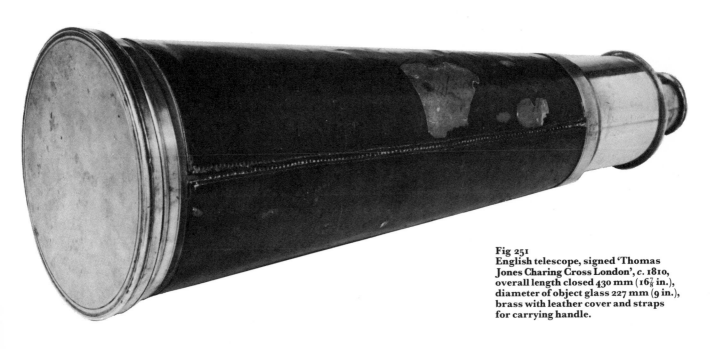

Fig 251
English telescope, signed 'Thomas
Jones Charing Cross London', c. 1810,
overall length closed 430 mm (16⅞ in.),
diameter of object glass 227 mm (9 in.),
brass with leather cover and straps
for carrying handle.

The Microscope

If the telescope revealed the facts about the physical world of the heavens which helped to revolutionize astronomical theory and the whole shape of Western thought, the microscope had equally fundamental consequences for man's ideas of scale, for it revealed a world of minute organisms such as had never before even been imagined. The name of Galileo is associated with this new instrument, as with the telescope. Galileo had adapted his refracting telescope to provide a form of compound microscope within months of making his first telescopic observations. In 1610 the Scotsman, John Wedderburn, a former pupil of Galileo's, mentioned his teacher's microscopic observations in a book published at Padua. Galileo's interest in the microscope seems to have continued at least until 1624.

The earliest work of natural history to show evidence of some form of microscope having been used, Francisco Stelluti's *Apiarum . . .*, was published in Rome in 1625. The earliest extant sketch (albeit crude) of a microscope is that which Isaac Beeckman drew in his journal 15 March 1631 to illustrate the instrument which Cornelius Drebbel had offered to James I in 1619. Drebbel was probably responsible for introducing the instrument to England, and an instrument like his was seen by Peiresc in Paris in 1622.

The manufacture of microscopes was slow, and it lagged behind the telescope in popularity. However, by 1654 the optician Johann Wiesel of Augsburg was making microscopes and the second half of the seventeenth century saw a great increase in their popularity. In England, especially after the publication of Robert Hooke's *Micrographia*, microscopy became a fashionable pursuit and a great boost was given to the manufacturing trade. By 1700 John Marshall had standardized lenses in semi-mass production. In Italy also, at the hands of Giuseppe Campani and Eustachio Divini, the microscope was greatly developed, popularized and reduced in price.

The purpose of the microscope is to make the viewing of minute objects possible. If, at a given distance, the angle subtended by an object falls below 1 minute of arc, then it ceases to be resolvable by the human eye (see fig. 253). By deploying one or more lenses, the microscope artificially increases the visual angle, this having the effect of allowing the eye to approach closer to the object. The instruments which perform this task are divided into the classes of *simple* or *compound*, according to whether they have one lens only, or a system of several.

The principle by which a simple microscope works is shown in fig. 252. A convex lens will act as a simple microscope in this way but with many imperfections. Small 'flea-glasses', comprising a short tube with a lens at one end and a clear glass plate at the other, on which an insect could be mounted for examination, were produced in the seventeenth century and later. Ultimately they found a new use in the twentieth century as a pocket viewer for photographic colour slides. The magnification obtained with them was low and an increase in power could only be obtained by increasing the radius of curvature until a full sphere was obtained. Thereafter further increase was gained by decreasing the size of the sphere. It was with lenses of this kind, and an ingenious mount of his own devising, that Leeuwenhoek made his fundamental observations on spermatozoa and other microscopic creatures between 1670 and c. 1720.

The eighteenth century saw several forms of mounting for simple microscopes, the Wilson screw-barrel type being particularly popular since it could convert easily into a small compound instrument (fig. 254–256). A further type of great popularity in the late eighteenth and early nineteenth centuries was the botanical microscope which received several slight adaptations by different makers (figs. 257–260). The more difficult innovation of jewel-lens instruments – microscopes in which the lens was made from a diamond or sapphire in order to reduce spherical aberration – was less popular, and only a small number were made, mainly between 1824 and 1837.

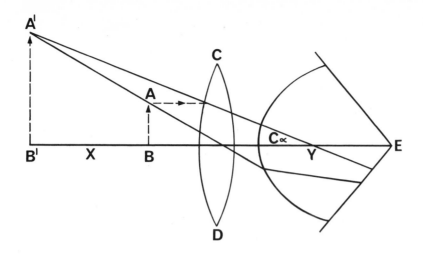

Fig 252
Action of the simple microscope.

XY is the least distance of distinct vision; AB is an object placed within it and the focal length of the lens CD. The image A¹B¹ appears erect and vertical and with an enlarged apparent visual angle.

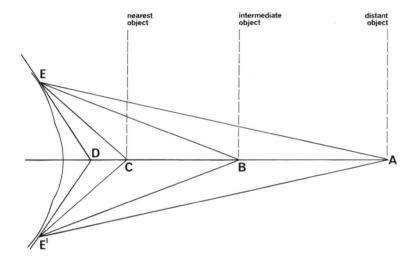

Fig 253
The relation of visual angle and distance.

As the angle subtended at the observer's eye EE¹ by the object placed successively at A, B and C increases, so the amount of detail which can be resolved by the eye is increased. At position B, however, the object falls within the least distance of distinct vision and is no longer resolvable.

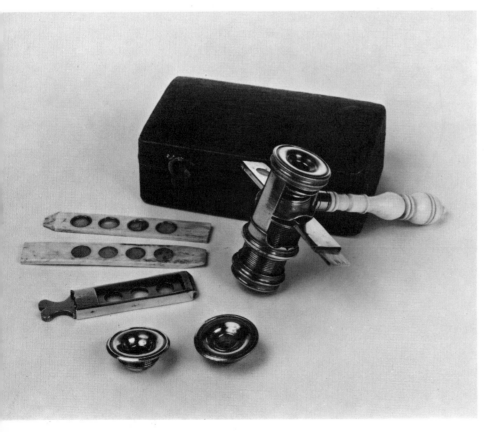

Fig 254
Wilson screw-barrel simple microscope, not signed, first half of eighteenth century, size of box 104 × 60 mm (4⅛ × 2⅜ in.), brass with ivory handle, green-velvet-lined shark-skin box.

Wilson's later form of screw-barrel, with turned handle, three sliders, one in an engraved brass mount, live object holder and three magnifiers.

Named after James Wilson, who published his version of the instrument developed by Nicholas Hartsoeker (c. 1694) in 1702, this small eminently portable instrument was one of the most popular among eighteenth-century amateur microscopists.

See James Wilson, 'The Description and manner of using a late invented set of small pocket microscopes, made by James Wilson', *Philosophical Transactions* 1702.

The action of the compound microscope (fig. 264) proceeds in two stages: first, an enlarged image of the object to be examined is formed; second, this image is magnified as if it were the object itself viewed through a simple microscope (fig. 261). The development of the compound microscope is essentially the history of increasing refinement in the optical performance of each of the components involved in this double action, and increasing precision and reliability in the mounting and motion-work of the instrument. Since microscopes are generally classified by form rather than by their optical fittings, it is to these that most attention is given in the descriptions. It should be stressed, however, that the real and basic developments in microscopical science came through a gradual improvement in precisely balanced compound lenses for eyepieces and object pieces. The varieties in style of mounting introduced by so many makers in the eighteenth century are relatively trivial. It is perhaps symptomatic of the greater attention paid to the optical side in the nineteenth century that mountings became less varied, but more solid and practical.

Fig 255
English simple microscope and Wilson screw-barrel microscope, signed 'E. Culpeper Londini', c. 1720, size of box 115 × 66 × 48 mm (4½ × 2½ × 1⅞ in.), ivory and brass, box of wood with velvet lining and fish-skin covering.

Wilson screw-barrel with turned handle and four numbered objectives; forceps in an engraved brass retention block with the initials 'EC'; five ivory slides; brass holder; talc box; simple microscope with engraved brass arm. In original fitted box.

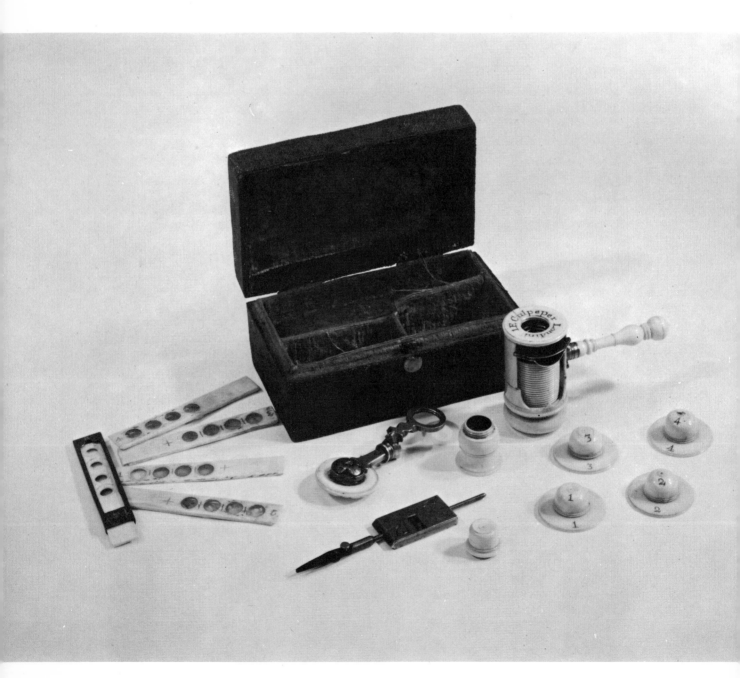

Fig 256

Wilson screw-barrel simple microscope, not signed, second half of eighteenth century, size of box 184 × 80 mm (7¼ × 3⅛ in.), brass, with green-velvet-lined fish-skin case, the wooden edges and rims painted red with brass catches and hinges.

Screw-barrel pocket microscope attached by a wing nut to a scroll stand which slots into the side of the case. Attached to the base of the stand by an horizontal arch is an adjustable mirror to enable the instrument to be used as a reflecting microscope. With brass tweezers, two ivory slides, turned ivory talc box and five numbered magnifiers.

See John Cuff *The Description of a new invention, to fix the pocket microscope, and make it answer the purposes of the large reflecting microscope* London, 1743.

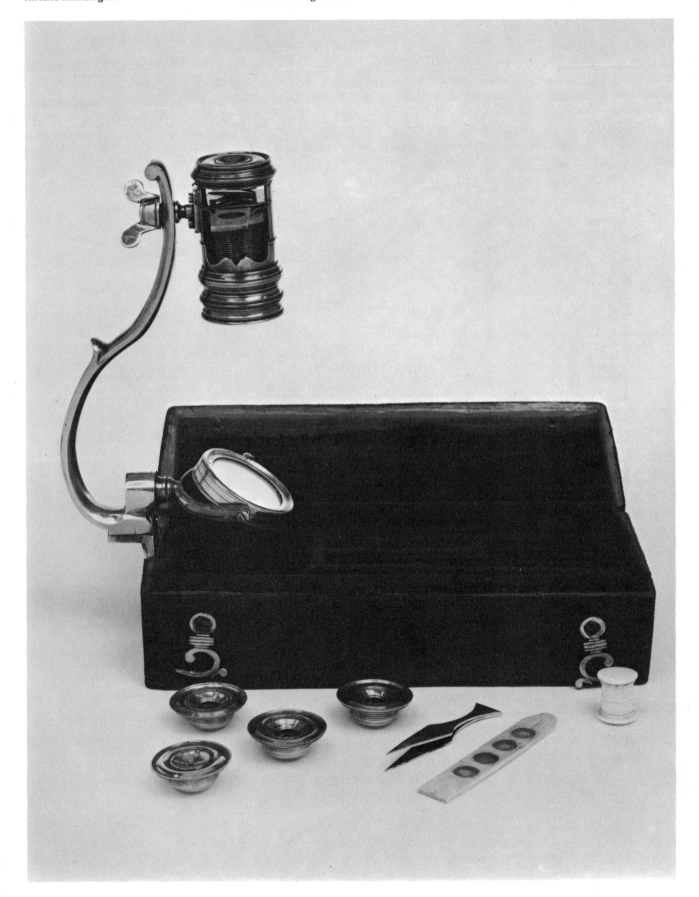

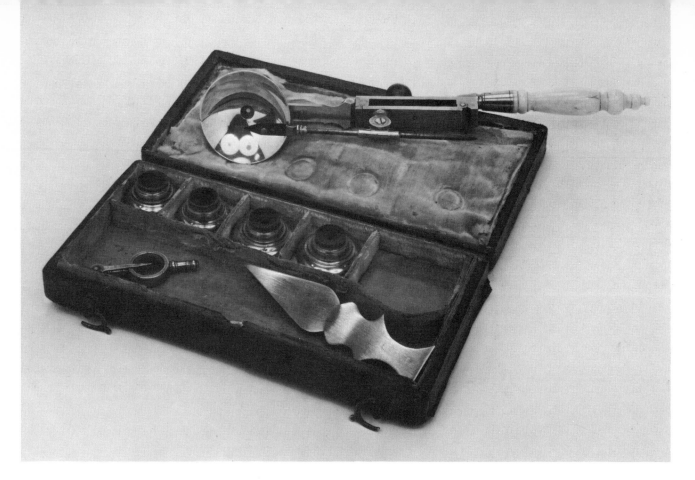

Fig 257
English compass microscope, not signed, eighteenth century, size of box 165 × 84 mm (6½ × 3¼ in.), brass and ivory, red velvet lined calf-skin case with brass catches and hinges.

Hinged to the brass frame is a spring-loaded brass strip which may be pressed away from the frame by the action of a screw. Mounted on the outer side of the strip is a steel wire having a point at one end and a pair of pincers at the other. Either end may be used for mounting specimens which will then be held in the centre of a 'Lieberkühn' screwed into the holder which is carried by an arm extended from the frame. A turned ivory handle screws into the lower end of the frame.

Fitted case and accessories, five 'Lieberkühns', brass tweezers, black/white objective plate.

Fig 258 *(below)*
English simple microscope, not signed, *c.* 1800, length 158 mm (6¼ in.) brass, ivory and lignum vitae.

Mounted on a turned handle, with a brass frame to which a pierced speculum is attached by a wing-nut. A spring-loaded ivory arm carrying the specimen-mount slots into a square opening at the base of the frame. On the upper part of the frame, a sliding brass bar containing two lenses is held by a spring, and passes across the aperture in the speculum.

Fig 259

English 'botanical' or aquatic microscope, signed 'D. ADAMS LONDON', early nineteenth century, size of box 190 × 144 mm (7½ × 5⅝ in.), brass with green-baize-lined mahogany box.

Cylindrical brass pillar mounted in a female screw on the instrument's mahogany box as pedestal. Inserted into the pillar is a rod adjustable by rack and pinion, with a transverse mounting at the top for the arm which carries the magnifiers. Attached to the cylindrical pillar is the spring stage with holes at the corners of the plate for attaching accessories. The reflecting mirror is placed towards the bottom of the pillar in a semi-circle on a horizontal arm which slots into the pillar.

The following accessories accompany the instrument: two 'Lieberkühns', three magnifiers, five ivory slides, two forceps, ivory talc box, pair of tweezers, folding ivory rule with brass joint, two ivory handled knives, two ivory handled probes.

A slightly modified form of the simple aquatic microscope named by Adams after John Ellis, who modified an instrument of John Cuff's for his own purposes in studying corallines and zoophytes. It was also recommended by William Curtis in his *Flora Londinensis*. Of the differences between the instrument illustrated by Adams and the present one, Adams noted in his text the use of rack-and-pinion adjustment and the insertion of the pin in the centre of the pillar rather than at the back.

See John Ellis *An Essay towards a natural history of the corallines and other marine products of the like kind* London, 1755.

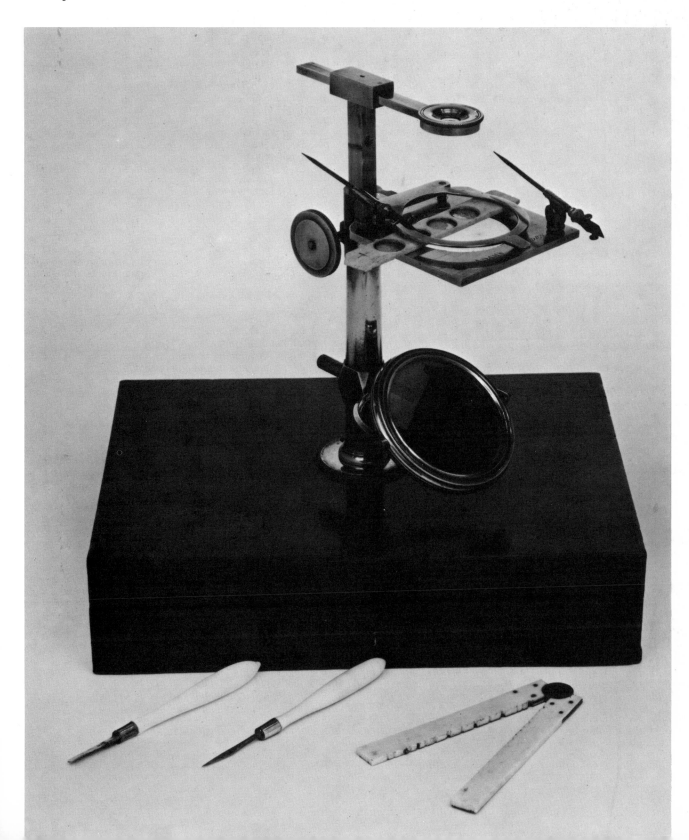

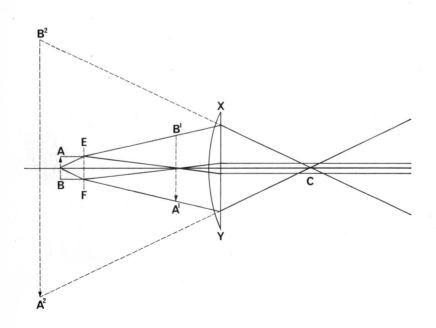

Fig 261
The principle of the compound microscope

An inverted image of the object is formed at A¹ B¹ by the object lens EF where it is magnified by the eye-lens XY to form a final image apparently placed at A² B².

Fig 260 *Centre*
English 'Botanic' microscope, not
signed, *c.*1790, size of box 120 × 63 mm
(4¾ × 2½ in.), brass, ivory and
mahogany.

Rectangular box with hinged lid and
spring catch. Mounted on the spindle
which supplies the hinge is a brass
pillar with a toothed edge against
which the stage may be adjusted by a
pinion. At the top of the column is an
ivory disc containing the lens. Both
lens-holder and stage are hinged to lie
flat when the box is closed. Attached
to the two inside edges of the box are
two sheaths of marbled paper, one
for forceps (now missing), the other
with a needle and knife, with ivory
handles. Pasted to the inside of the
lid is a printed paper containing
instructions for use.

This form of microscope was
designed by William Withering for
botanical use, and described in the
second and subsequent editions of his
*Botanical arrangement of British
plants.*

Right
English 'entomological' microscope,
not signed, *c.* 1790.

Fig 262
English compound microscope,
signed 'I. MARSHALL', between *c.*1700
and *c.*1725, overall height 400 mm
(15¾ in.), gold tooled green morocco
barrel, lignum vitae, walnut and oak.

Octagonal box with accessories
drawer carrying sub-stage mirror
(? a slightly later replacement) and
mount for the universal ball joint
carrying the pillar marked with six
setting positions. Fine adjustment is
effected by an endless screw. Body
tube with tooling characteristic for
the period 1700–1725.

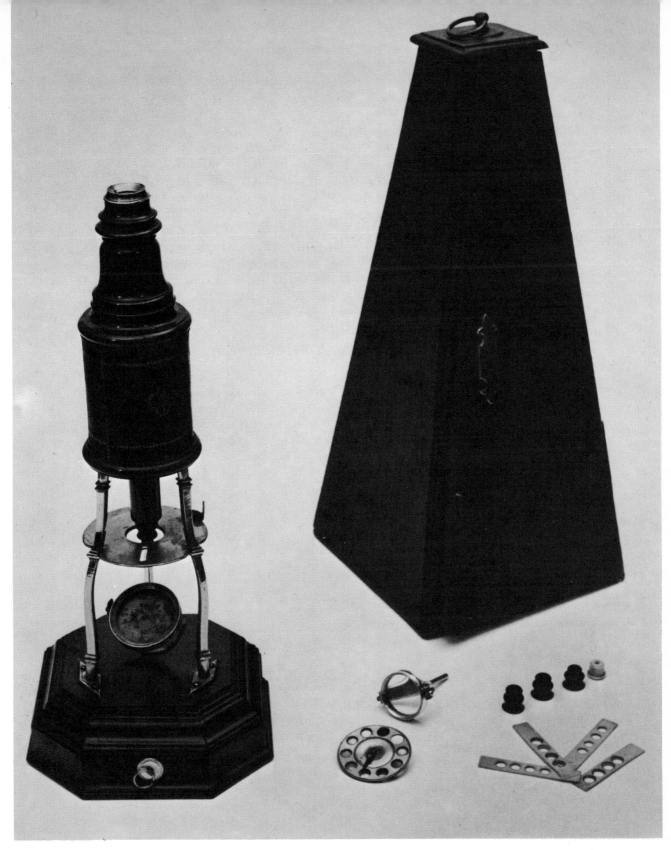

Fig 263 *left*
English compound microscope,
signed on label in drawer 'Matthew
Loft Maker at the Golden spectacles
the Backside of the Royal Exchange
LONDON', *c.*1720, height 400 mm (15¾ in.),
brass, ray-skin and oak.

Culpeper-type microscope, octagonal
base with accessory drawer on which
are mounted the sub-stage mirror and
the three diamond-section curved legs
carrying the stage and the body. The
accessories include: eight slides, five
objectives, spring stage, fish plate,
revolving object carrier, live box,
forceps, glass tube, talc box.

Fig 264
English Culpeper-type compound
microscope, first half of eighteenth
century, not signed, height 450 mm
(17¾ in.), body of lignum vitae, black
leather and cardboard, with gold-
tooled decoration, brass legs and
stage, mahogany base. Oak case with
engraved brass keyhole, brass
carrying handle and hinges.

Octagonal base, with accessory
drawer, on which are mounted the
three square-section continuous bent
legs which support the body and carry
the stage. Concave reflecting mirror
mounted on the pedestal below the
stage aperture. Draw-tube focusing.

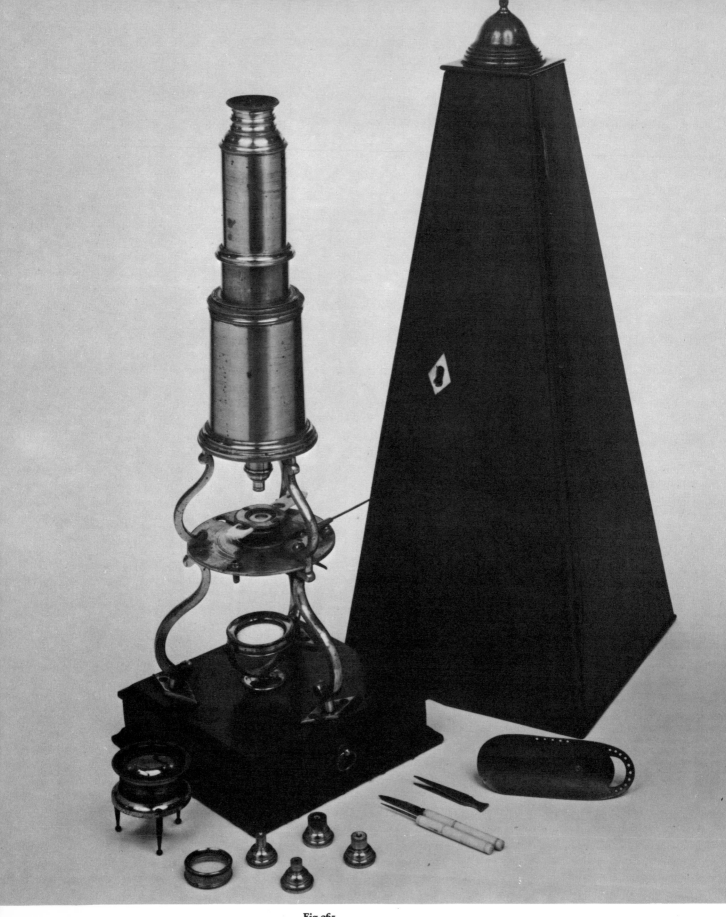

Fig 265
**English Culpeper-type compound
microscope, not signed, *c.*1800, overall
height of box 445 mm (17½ in.), brass on
mahogany stand with mahogany box
with inlaid ivory key-hole.**

**The body of the instrument, with the
inner tube sliding for focus, is
mounted by three scroll legs on the
stage, itself supported by a further
three scroll legs which are screwed to**

**the pedestal. Screwed to the centre of
the pedestal below the stage aperture
is a concave reflecting mirror.**

**The following accessories are
contained in the drawer in the
pedestal: five magnifiers, five ivory
slides, lens, fish-plate, forceps, ivory-
handled knife with steel blade, ivory-
handled probe with steel needle, live
specimen box.**

Fig 266
German compound microscope, not
signed, c.1800, overall height 295 mm
(11⅝ in.), lightwood and simulated
rayskin.

Turned circular base with cushion
feet carrying an obliquely mounted
mirror; stage supported on three
straight turned legs. Outer tube of
covered cardboard attached to the
stage which incorporates a Bonnani
spring. Draw-tube of wood covered in
paper. Body tube of wood partly paper
covered, with screw-ended nosepiece.

Fig 267
German compound microscope, not
signed, late eighteenth century,
softwood, paper, hardwood and
simulated fish-skin.

Simple draw-tube microscopes of this
kind were produced in great numbers
by eighteenth- and early-nineteenth-
century toymakers in Nuremburg.

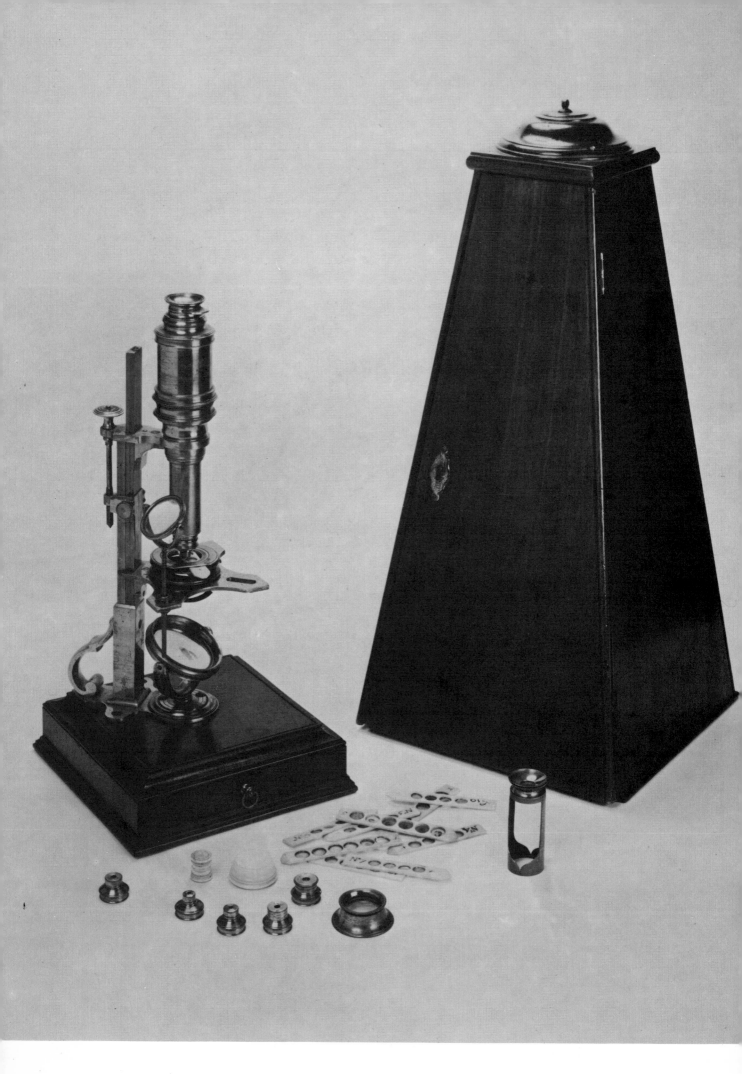

Fig 268
English compound microscope,
signed 'STERROP LONDON', *c.*1780, height
440 mm (17¼ in.), brass, with
mahogany case.

Cuff-type microscope, the body
attached by an horizontal area (from
which it may be removed) to a pillar
sliding over the fixed pillar (marked
1–6) mounted on the base with shaped
buttress. Attached to the collars
circling the pillars and housing the
lower end of the adjusting screw is a
clamping screw for fixing the
instrument when one of the numbered
positions (which correspond to the
numbers of the magnifiers) has been
adopted. The spring stage is attached
to the fixed pillar and carries on one
corner an adjustable and removable
condenser lens. A slit in a second
corner of the stage allows for the
attachment of a fish-plate. The
adjustable concave mirror for
reflecting light is mounted directly
below the stage on the pedestal. The
whole instrument is mounted on a
mahogany pedestal with accessory
drawer. The accessories include
'Lieberkühn' holder, nine slides, talc
box and alternate lenses.

One of the most popular forms of
microscope in the later eighteenth
century a description of this
instrument was published by John
Cuff in 1744 in *The Description of a
new constructed double microscope,
in which some useful improvements
are introduced.*

Fig 269
English 'Universal' microscope, not
signed but by Francis Watkins,
c. 1754, size of box 236 × 167 mm
(9¼ × 6½ in.), brass, mahogany box
with brass clasps, hinges and
carrying handles.

*The simple compound
microscope*
Brass stand to which is screwed a
folding tripod, carrying at its top end
an adjustable arm to which are
attached the pillar and the
independently adjustable plano-
concave mirror. Each side of the pillar
is numbered 1–7 corresponding to the
magnification powers, the side
marked S for use when the instrument
is set up in simple form, and the side D
for double or compound form.
Attached to the pillar on a sliding
bracket is a spring stage, fine
adjustment being effected by the
screw at the base of the pillar. When
used in its simple form, a double
brass plate containing between the
plates a rotatable disc with seven
numbered lenses is attached to the
arm at the top end of the pillar. When
used in its compound form, as
illustrated, the tapered barrel is
screwed to the arm and the D
magnifier scale is used.

The solar microscope
In a fitted fish-skin box 185 × 153 mm
(7½ × 6⅛ in.), slotting inside the
mahogany case, is a brass common
solar microscope consisting of two
body tubes mounted on an adjustable

circular disc in the square mounting
plate, to which is attached on the
opposite side a rectangular sunlight-
collecting mirror. A single microscope
with screw focusing is attached to the
end of the second body tube.

Accessories
Box of twelve sliders, six with objects
ready mounted, six plain. Brass slider
for liquid preparations, forceps, three
Lieberkühns, black/white cylinders,
tweezers, fish-plate, brush, spiral
cotton-holder, folding brass live
specimen holder, aquatic box. For a
detailed description of this form of
microscope see Francis Watkins
*L'Exercise du microscope … la
description d'un microscope, qu'on
peut appeler universe, d'autant qu'on
y trouve les propriétés de toutes les
differentes sortes qui ayant encore
parues, construit sur un nouveau
plan* London, 1754.

Behind the solar microscope is a copy
of John Evelyn *Silva, or a discourse of
forest trees, and the propogation of
timber in his Majesties dominions …*
with notes by A. Hunter, third edition
(of the re-issue), York, 1801 with Henry
Cavendish's signature 'H. Cavendish
1802' on the top right corner of the title
page.

Fig 270 *right*
Common solar microscope, signed
'RAMSDEN LONDON', late eighteenth
century, size of box 250 × 140 mm
(10 × 5¾ in.), brass.

The double-tube body of the
instrument screws at one end into a
circular plate screwed on to the
square mounting plate. Attached to
this circular plate is a rectangular
frame containing an adjustable plane
mirror for collecting sunlight which
is then gathered by the lens set in the
end of the body. At the other end of the
body a microscope with rack-and-
pinion focusing is screwed on. In use
the square plate of the instrument is
screwed to a shutter with the
rectangular mirror passing through
a hole to the outside. By adjusting the
position of the mirror the sun's rays
can be focused through the
instrument to form an image of
specimens of the object under
examination on a screen conveniently
placed in a darkened room.

Fig 271
English Universal compound
microscope, signed 'B. Martin Inv'. et
Fecit, London', size of box 285 ×
201 mm (11¼ × 7⅞ in.) brass, with
green-velvet-lined fish-skin covered
case, the wooden sides and edges
painted red.

Tripod stand carrying fixed
perpendicular triangular-shaped
pillar with turned base. The double-
sided mirror-holder is mounted on
a curved bracket attached to a collar
sliding on the pillar. The stage is
attached to a similar collar adjustable
by rack and pinion; screwed to the
top of the pillar and rotatable (but
now clamped by a pin passing
through the mount) is a right-angled
bracket to the top of which is hinged
an arm into which the body of the
instrument screws. The body has
Martin's 'between lens' and rack-and-
pinion adjustment. When the hinge is
lifted up, a holder into which a single
lens, or 'Lieberkühn', can be screwed,
may be inserted into a slot in the
bracket, thus converting the
instrument for use as a simple
microscope.

The following accessories are
contained in the fitted case: eight
magnifiers, two Lieberkühns, live
specimen box, forceps, brass slider
in holder, fish-pan, multiple holder
for stage, brass mount for compound
sub-stage condenser, plane glasses for
stage. Sharing several design features
with Martin's 'Grand Universal'
microscope, the present instrument
seems to be a transitional model
between this and his preceding 'New
Universal'.

Fig 272
English compound microscope, signed 'Fraser Bond St LONDON', c. 1780, size of box 173 × 121 mm (6⅞ × 4¾ in.), brass, with green-velvet-lined fish-skin box.

The accessories include five objectives, 'Lieberkühn', tweezers, needle, probe, sliders, mirror, condenser lens, black/white plate and forceps.

Fig 273
English compound microscope, not signed, c. 1800, brass with mahogany box.

Tripod stand carrying a fixed square section pillar, to which is attached the square stage adjustable by a screw thread. The body of the instrument screws into an horizontal arm attached to the top of the pillar. The double-sided mirror is attached horizontally to the base of the pillar.

With accessories in fitted box: two Lieberkühns, six magnifiers, forceps, fish-pan, five wood sliders for opaque objects, six boxwood sliders, twenty glass slips, aquatic box, hand magnifier brass turpentine or gum box, brass rimmed plane glass, glass phial, stage glasses.

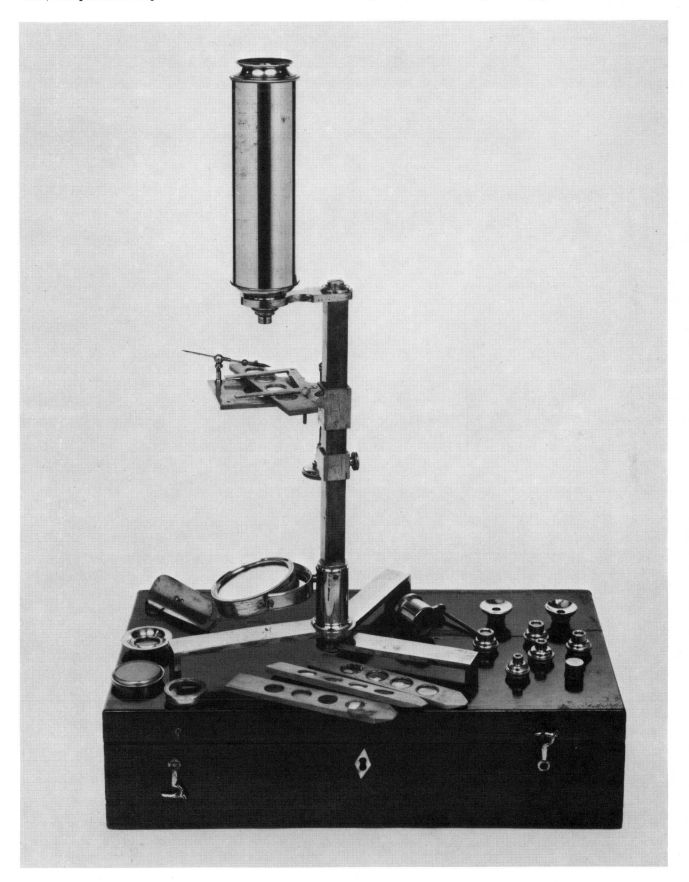

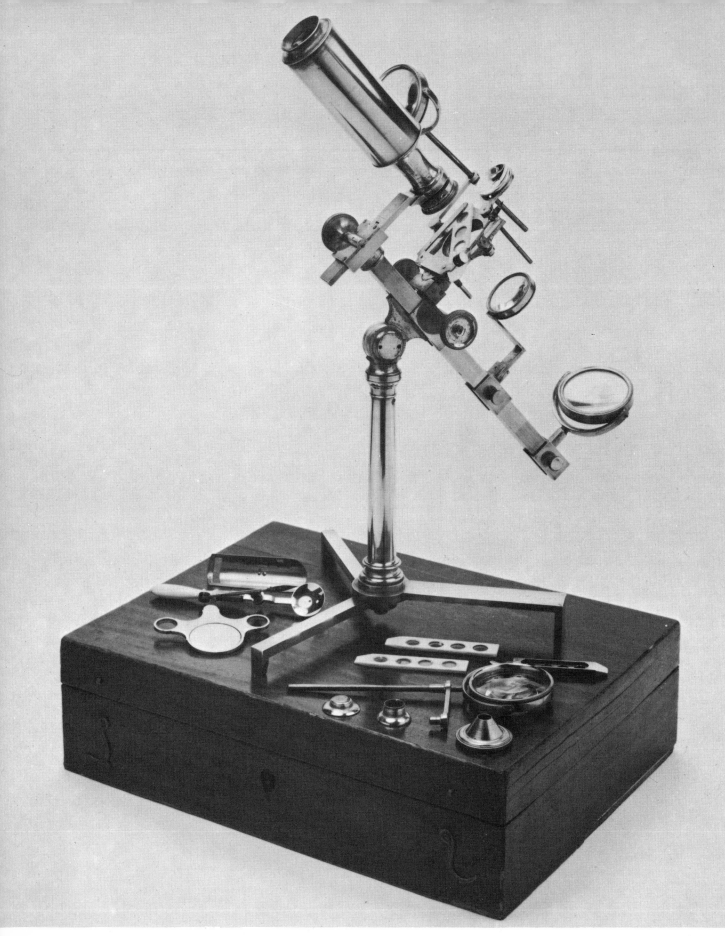

Fig 274
English compound microscope,
signed 'Gilbert and Sons, LONDON',
c. 1810–20, size of box 266 × 365 mm
(10½ × 14⅜ in.), brass with mahogany
box.

Flat tripod foot with folding feet from
which arises a circular section pillar
having at the top a compass joint. To
this joint a square brass bar is
mounted carrying the mirror and
condenser holder on separate sliding
fittings provided with clasps. Square-
type stage with rackwork focusing,
the rack being let into the bar serving
as pillar. The body is mounted on a
horizontal arm, with rackwork for
motion vertically above the stage, at
the top of the pillar. Accessories
include, fish-plate, forceps, sliders,
condenser, 'Lieberkühn', and fitted
box.
An example of the 'Jones Most
Improved' type of microscope
developed by W. & S. Jones at the end
of the eighteenth century. The 'Jones'
instrument was based on the Cuff-
type microscope but incorporated
various other improvements.

Fig 275
Carpenter's improved compound microscope, not signed, *c.*1830, case 397 × 173 mm (15⅝ × 6¾ in.), brass with purple-velvet-lined mahogany case.

Tripod stand which carries the cylindrical pillar on a compass joint; fixed double spring stage and detachable concave mirror. The body, with rack and pinion focusing is held on a fixed arm attached to the top of the pillar.

In fitted box with inset brass name-plate (blank) with accessories. Seven numbered magnifiers (0–6),

Lieberkühn, hand magnifier aquatic box, brass turpentine or gum box, brass talc box, brass-rimmed plane glass for water-plants, two glass slips for examining crystalization, brass tweezers, condenser lens on stand with candle holder, six wood slides with opaque objects mounted on pasteboard, six ivory slides, two large boxwood slides, forceps, cardboard shade.

Also included is a copy of *A Description of the new improved compound microscope for opaque and transparent objects* (n.d., n.p.), in which ascription of this model microscope is made to Carpenter.

Fig 276
Cary-type compound and single pocket microscope, signed 'Cary, LONDON' *c.*1820, 95 × 78 mm (3¾ × 3¹⁄₁₆ in.), brass, with mahogany box.

Conical brass body containing eyeglass and middle-glass, attached to the pillar by a sliding bracket. Spring stage mounted on pillar with rack and pinion adjustment. Mounted on the rim of the box by a female screw placed in the position of the lock.

Accessories: ivory slider, brass tweezers, aquatic box, three magnifiers, forceps.

Fig 277
German photographer's lamp, signed
'E. LEITZ WETZLER', nineteenth century,
410 mm (16¼ in.), mahogany, brass,
and sheet metal.

Micrometers and eyepieces

Eyepieces of telescopes or microscopes vary in construction according to the purpose for which they are to be used. In low-power instruments a single lens may be used, but as magnification is increased so imperfections are magnified. In the simplest form of compound eyepiece, a field lens is placed in front of the eye-glass in order to concentrate the rays from the object-glass into a smaller image. This is viewed through the eye-glass. Since by this means the whole of the image is seen and no light is lost, the image appears brighter than it would were the field lens removed.

For a distinct and well-defined image in a flat field, however, probably the best eyepiece was that devised by Christian Huygens and named after him (fig. 278). This consists of two plano-convex lenses, one having a focal length two or three times greater than the other. This larger glass supplies the place of the field lens which intercepts the light rays from the object, and brings them to a focus more quickly. Extraneous light is shut out by a diaphragm and the image formed by the field lens is viewed through the eye-glass. The arrangement appreciably reduces spherical and chromatic aberration if properly adjusted and produces an inverted image. Ramsden devised a second form of celestial eyepiece producing an inverted image which became commonly used particularly for telescopes to which micrometers were to be fitted (fig. 279). This employed a convexo-plane and a plano-convex lens with their curved surfaces turned towards each other. Further improvements to it were made by Kellner and J. H. Steinheil.

A basic problem for the astronomer is measuring the size of an object which he views and determining its position. For the latter purpose cross-wires or spider-lines were placed at the point of the image in the field of view. For the former various forms of micrometer were developed. Both innovations are associated with the name of William Gascoigne (?1612–44) who was one of a small group of dedicated scientists in Lancashire in the 1640s and 50s. Following his initiative numerous types of micrometer were invented although very few examples survive from the seventeenth century. In the eighteenth, however, much attention was paid to this

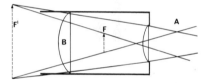

Fig 278
Huygen's eyepiece

Two plano-convex lenses A and B have their convex sides turned towards the object glass, and have focal lengths in the proportion of 3:1. The weaker lens, B, placed inside the focus of the object glass brings the image to a focus at F where it is viewed by the eye-lens A. A magnified image appears to be seen at F¹.

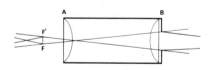

Fig 279
Ramsden's eyepiece

A and B are two plano-convex lenses of the same focus which are placed apart at a distance equal to two thirds of the focal length of either, on the eye side of the focus of the telescope. They act together to render the rays parallel and produce a magnified inverted image FF¹.

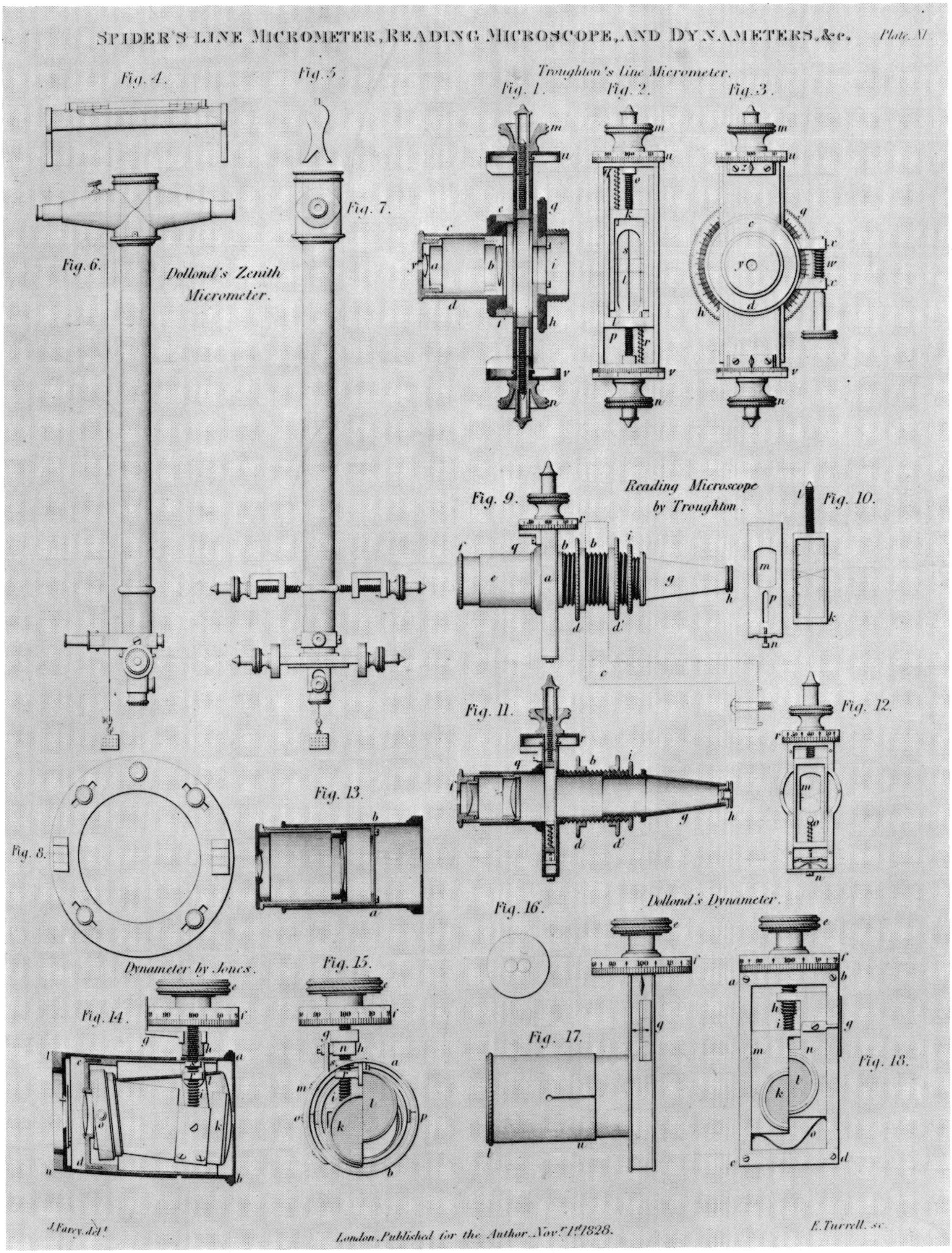
SPIDER'S LINE MICROMETER, READING MICROSCOPE, AND DYNAMETERS, &c.
Plate XI.
Troughton's line Micrometer.
Fig. 1.
Fig. 2.
Fig. 3.
Fig. 4.
Fig. 5.
Fig. 6.
Fig. 7.
Dollond's Zenith Micrometer.
Reading Microscope by Troughton.
Fig. 9.
Fig. 10.
Fig. 11.
Fig. 12.
Fig. 8.
Fig. 13.
Fig. 16.
Dollond's Dynameter.
Dynameter by Jones.
Fig. 14.
Fig. 15.
Fig. 17.
Fig. 18.
J. Farey del.
London, Published for the Author, Nov. 1st 1828.
E. Turrell sc.

Fig 280
Micrometers, from W. Pearson *An Introduction to practical astronomy containing descriptions of the various instruments, that have been usefully employed in determining the places of the heavenly bodies with an account of the methods of adjusting and using them* London, 1829, vol ii, plate xi.

device, the improvements made by John Dollond to the divided object glass micrometer being particularly important. By the time that William Pearson (1767–1848) came to survey these measuring accessories in his *Introduction to Practical Astronomy* (fig. 280) he was able to discuss an impressive range of instruments.

Other optical instruments

The behaviour of light and the phenomena of vision have exerted a long fascination on mankind, both as serious subjects of scientific research, and as pleasant pastimes and recreation. Many of the instruments of entertainment and experiment still survive and may still delight their possessors. Of all perhaps the burning glass has the longest continuous history, but few surviving examples are likely to be much older than the eighteenth century. Then, as gentlemen amateurs repeated the experiments of Sir Isaac Newton and optics became fashionable, a great deal of apparatus was constructed. The burning glass was an essential, if fragile, piece of laboratory equipment. Its most famous use perhaps was in the experiments of solar heat which led Sir William Herschel to the discovery of infra-red rays. Continuing in use throughout the nineteenth century, good examples are still occasionally to be found.

With the emergence of a class of professional laboratory scientists in the nineteenth century a manufacturing industry developed out of the older class of mathematical practitioners to meet their need for equipment. Among the many devices developed several, such as polarimeters, for measuring the amount of polarized light in a beam, goniometers, for measuring the refracting surfaces of crystals, and spectroscopes, were optical, or optically based. Many of these devices survive, often lurking unrecognized in attics and tool-chests, and are fine examples of the

Fig 281
English camera obscura, not signed, *c.* 1710, wood, gold-tooled vellum-covered tube to lens mount, mirror, the ground glass screen is missing.

The tooling on the lens mount is similar to that found on telescope tubes of the period *c.* 1710–25 and is possibly part of an adapted telescope tube.

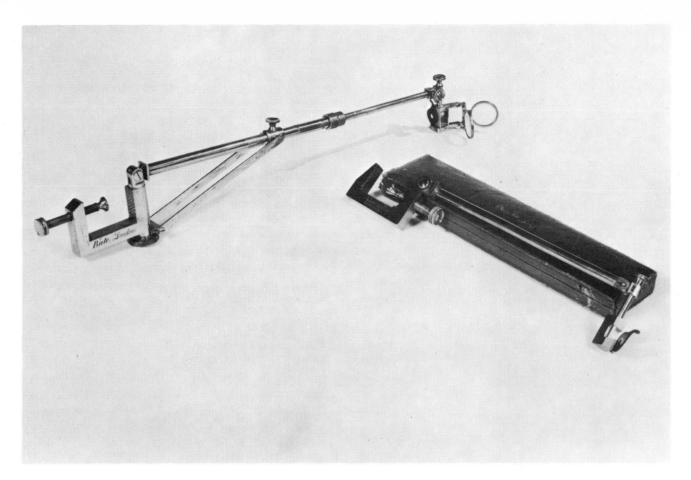

Fig 282
English cameras lucidae

Right
Signed 'BANCKS LONDON', *c.* 1810, 242 mm
(9½ in.), brass.
Left
Signed 'Bate London', *c.* 1830, 373 mm
(14¾ in.), brass.

A prism is attached by an axle parallel
to its edges to the end of a rod sliding
in a tube, which has, at its other end, a
clamp for fixing to the edge of a table.
The tube is attached to the clamp by a
knuckle-joint. A setting piece by
which the tube may be fixed at any
desired inclination is also jointed to
the clamp, acting on the tube by a
collar and clamping screw.

Fig 283
Scioptic ball, not signed, eighteenth
century, overall diameter 142 mm
(5½ in.), lignum vitae and Spanish
mahogany.

Two lenses are mounted in the lignum
sphere which is adjustable in a socket
and frame.

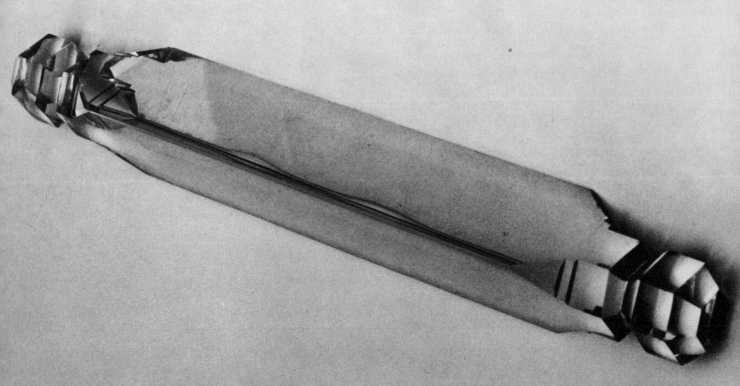

Fig 284
English stereoscope, not signed, late
nineteenth century, 335 × 185 mm
(13¼ × 7¼ in.), mahogany.

Moulded base with inclinable viewing
platform. Fretted holding piece.

precision instrument-maker's craft. Optical principles were also put to use for other purposes in instruments such as the camera obscura (fig. 281) and camera lucida (fig. 282) to provide an easy means of accurate copying. Other instruments were also developed for this purpose during the nineteenth century, such as the 'graphic telescope' of Cornelius Varley.

The camera obscura, which dates from at least the eleventh century, could be used in two forms. In what was perhaps its earliest manifestation a hole was cut in the shutter of a window or a wall, and a convex lens inserted in it. If the rest of the room was then darkened and a sheet of paper placed in the focus of the lens, an image of the exterior would be formed on the sheet. If this were fixed down firmly, the outline could then be drawn round. From the mid-seventeenth century onwards, if a whole room was intended to be used as a camera, a scioptic ball might be fixed to a window shutter (fig. 283). The first suggestions for the second form of camera obscura, which was portable for taking landscapes and military surveys, appear to date from the end of the sixteenth century, and in the following centuries various types were suggested and made. Being made of perishable materials they are now somewhat rare. It was, however, this instrument which led to attempts to find by chemical means some way of fixing permanently upon its paper the image cast by the camera obscura. Investigations to this end were carried out by several men throughout Europe during the eighteenth century, but it was not until 1826 that Joseph Nicéphore Niepce (1765–1833) succeeded in producing a permanently fixed picture, using a camera obscura made by the Paris opticians Charles and Vincent Chevalier. Subsequent development at the hands of Louis Jacques Mondé Daguerre (1787–1851) and William Henry Fox Talbot (1800–77) quickly followed by Sir John Herschel (1792–1871) refined and made practical the processes involved. In 1839 the first cameras were advertised by Francis West, an optician of Fleet Street.

The great success of photography in the nineteenth century was largely caused by its entertainment value. To a similar end the optical instrument-makers devoted much energy and ingenuity in the production of a wide range of toys based on indirect, distorted and enlarged vision, or in moving images. Although not themselves scientific instruments, these devices should not be overlooked. Some of them, such as the anamorphoscope (a cylindrical mirror for viewing a distorted picture), zograscope and stereoscope have long histories, while the moving image devices such as the zoetrope, praxinoscope and kinorama are among the forerunners of the cinematograph. It was the sale of such toys as these which helped to capitalize the more difficult research work of nineteenth century optical makers, and in this way the toys played a part in the history of science. They may surely challenge for a place in any collection of scientific instruments.

Fig 285
Prism, nineteenth century, not signed,
overall length 306 mm (12⅛ in.), width
of face 47 mm (1⅞ in.), glass with
moulded handles.

GLOSSARY OF TERMS

Terms not listed here are described in the text and may be located through the index.

Alidade — A sighting device attached to an astrolabe, quadrant or other graduated instrument, movable about the centre of a graduated arc, and which indicates the number of degrees cut off on the limb of the instrument.

Alt-azimuth — An instrument for determining altitudes and azimuths.

Annulus — A ring.

Aspectarium — A diagram showing the positions of the planets in relation to each other.

Azimuth — The angular distance measured along the horizon between the meridian of a place and the vertical circle passing through the centre of a celestial object and the zenith.

Colure — Either of the two great circles supposed to intersect each other at right angles in the poles of the equator. The two circles pass respectively through the equinoxes and the solstices.

Ecliptic — The path which the sun, owing to the annual revolution of the earth, appears to describe among the stars. It makes an angle with the equinoctial of about $23°27'$.

Epact — The position of the year in a nineteen-year cycle which begins when the new moon apparently falls on the 1 January of the solar year, and at the end of the cycle does so once more.

Ephemerides — A table showing the computed positions of a heavenly body for a given period.

Equinox — The time when the sun reaches one of the two equinoctial points, or points in which the ecliptic and the celestial equator intersect each other. The vernal equinox falls about 21 March, the autumnal about 23 September. These are the only two occasions in the year when day and night are of equal length throughout the world.

Fiducial — A line or point assumed as a fixed basis for comparison.

Gimbals	Two rings moving on pivots so as to have a free motion in two directions at right angles and thus to keep steady any object they carry. Primarily used at sea for chronometers and compasses.
Gnomon	The shadow-casting element in a sundial.
Great circle	Any circle on the celestial sphere passing through a given point and the pole.
Lieberkühn	A lens surrounded by a concave reflector fixed to the object-end of a microscope to bring light to a focus on an opaque specimen.
Loxodrome line	Lines of oblique sailing.
Meridian	Any great circle passing through the zenith and the poles.
Planisphere	A projection of a sphere onto a flat surface.
Precession	The successively earlier occurrence of the equinoxes in each sidereal year, caused by their retrograde backward motion along the ecliptic produced by the slow change of direction in space of the earth's axis.
Rutter	A book of marine routes and sailing directions.
Solstices	The points on the ecliptic and time of year when the sun is at its greatest distance north and south of the equator; i.e., at midsummer, 21 June, and midwinter 22 December.
Vernier	A short scale movable against the graduated scales of measuring instruments by which more minute divisions may be read off.
Volvelle	One or more movable discs carrying, or surrounded by, graduated or figured circles.

FURTHER READING

Good general books on scientific instruments are few. Among the more useful are Henri Michel *Scientific Instruments in Art and History* translated by Francis R. and R. E. W. Maddison, London and New York, 1967; Francis Maddison *A Supplement to a Catalogue of Scientific Instruments in the Collection of J. A. Billmeir, Esq., C.B.E.* London and Oxford 1957; Maurice Daumas *Scientific Instruments in the Seventeenth and Eighteenth Centuries* translated by Mary Holbrook, London 1973. The first two volumes of R. T. Gunther *Early Science in Oxford* Oxford 1921–3 also retain some value. For the general background and biographical information about English instrument-makers, E. G. R. Taylor *The Mathematical Practitioners of Tudor and Stuart England* Cambridge 1967 and New York 1966, and *The Mathematical Practitioners of Hanovarian England* Cambridge and New York 1966 are essential but must be used with great caution. The basic bibliographical guides are Francis Maddison, 'Early Astronomical and Mathematical Instruments. A Brief Survey of Sources and Modern Studies', *History of Science* ii, 1963, pp. 17–50, and G. L. 'E Turner, 'The History of Optical Instruments. A Brief Survey of Sources and Modern Studies', *History of Science* viii, 1969, pp. 53–93. The only worthwhile account of eighteenth-century physical apparatus is provided by G. L. 'E Turner, 'Descriptive catalogue of Van Marum's scientific Instruments in Teyler's Museum' in E. Lefebvre & J. G. de Bruyn (ed.) *Martinus van Marum, Life and Work* vol iv, Leyden 1973, pp. 127–397 (also issued separately).

Astronomy
For the background of changing astronomical theories see J. L. E. Dreyer *History of Planetary Theories from Thales to Kepler* London 1906. On the astrolabe the classic work is that of Henri Michel *Traité de l'astrolabe* Paris 1947, of which a revised edition with English translation is in preparation. Brief accounts of the instrument in English are Willy Hartner, 'Asturlab' and 'The Principles and use of the astrolabe' in *Oriens Occidens* Hildesheim 1968; J. D. North, 'The Astrolabe' *Scientific American* vol. 230, 1 January 1974. For universal astrolabes see Michel, and Francis Maddison, 'Hugo Helt and the Rojas astrolabe projection', *Agrupamento de estudos cartografia antiga* xii, Coimbra 1966. The standard work on globes is E. L. Stevenson *Terrestrial and Celestial Globes* 2 vols., New Haven and London, 1921. For the orrery see Howard C. Rice *The Rittenhouse Orrery; Princeton's Eighteenth Century Planetarium, 1764* Princeton 1954 and for a more general background, A. J. Turner *The Clockwork of the Heavens* London, 1973. Zdeněk Horský and Otilie Škopova *Astronomy Gnomonics* Prague 1968, illustrate several attractive instruments, while for German instruments Ernst Zinner *Deutsches und Niederländische astronomische Instrumente des 11–18 Jahrhunderts* Munich 1956, is essential.

Navigation

For general accounts of the development of navigation see E. G. R. Taylor *The Haven Finding Art* second edition, London 1971; E. G. R. Taylor and M. W. Richey, *The Geometrical Seaman, a book of early Nautical Instruments* London 1972 and New York 1963; W. E. May *A History of Marine Navigation* Henley-on-Thames 1973. D. W. Waters *The Art of Navigation in England in Elizabethan and Early Stuart Times* London 1968, is the only detailed treatment of one period. H. O. Hill and E. W. Paget-Tomlinson *Instruments of Navigation* London 1958, is useful out of all proportion to its size. On the mariner's astrolabe see R. C. W. Anderson *The Mariner's Astrolabe* Edinburgh 1972.

Sundials

For sundials, see Kathleen Higgins, 'The Classification of sundials', *Annals of Science* ix, 4, 1953 pp. 342–58; Henri Michel *Les Cadrans solaires de Max Elskamp* Liège 1966; René R. J. Rohr *Sundials, History, Theory and Practise* Toronto 1970.

Surveying

Surveying instruments have been less well studied than other groups. E. R. Kiely *Surveying Instruments, Their History and Classroom Use* New York 1947 and A. W. Richeson *English Land Measuring to 1800, Instruments and Practise* Cambridge Mass. and London 1968 offer introductory guides. A short account of urban surveying is given by D. J. Bryden in Phillipa Glanville *London in Maps* London 1971. For nineteenth-century instruments, William Ford Stanley, *Surveying and Levelling Instruments Theoretically and Practically Described* fourth edition, London 1914, supplies a comprehensive introduction.

Optics

The bibliography of optics is huge. For spectacles see Richard Corson *Fashions in Eyeglasses* Chester Springs Pa. 1967 and London 1968. On the telescope H. C. King, *The History of the Telescope* London 1955 and for Galileo and the telescope, Francis Maddison, 'Galileo and the history of scientific instruments' in *Saggi su Galileo* Pisa (forthcoming). On the microscope, Reginald S. Clay and Thomas H. Court *The History of the Microscope* London 1932 should be used cautiously in the light of G. L. 'E Turner's comments in 'Micrographia Historica: the Study of the History of the Microscope', *Proceedings of the Royal Microscopical Society* vii, 2, 1972, pp. 1220–49. More valuable is Alfred N. Disney, Cyril L. Hill and Wilfred E. Watson Baker, *Origin and Development of the Microscope* London 1928. A short modern survey is provided by S. Bradbury *The Microscope Past and Present* Oxford and Elmsford N.Y. 1968, an abridged and expanded edition of the same author's *The Evolution of the Microscope* Oxford 1967. For photographic apparatus see the excellent monograph by D. B. Thomas *The Science Museum Photography Collection* London 1969.

ACKNOWLEDGEMENTS FOR ILLUSTRATIONS

Frontispiece and figures 21, 186 courtesy of the Time Museum, Rockford. Figures 6, 60, 61, 74, 80, 92 Crown copyright, reproduced by permission of the trustees of the National Maritime Museum. Figures 36, 37, 192, 223 courtesy of Alain Brieux. Figures 20, 22 courtesy of the Rijksmuseum, Nederlands Scheepvaart Museum. Figure 47 from the collection of H. A. L. Dawes. Figure 54 from the collection of Gilbert Suc. Figures 58, 64, 65, 103, 106, 114, 115, 123, 129, 148, 155, 156, 163, 165, 167, 175, 176, 180, 188, 206 by courtesy of Sotheby & Co. Figures 77, 83, 94, 164, 166, 185, 245, 277 by courtesy of Christie, Manson and Wood. Figure 84 by permission Dundee Museums and Art Galleries. Figure 101 the Hull Fisheries Museum. Figure 105 Asprey & Co. Figure 81 from the collection of I. F. W. Devereux. Figures 108, 208, 218 Nicolas E. Landau. Figures 120, 135, 136, 147 from the collection of Christopher Sykes at Woburn. Figures 63, 159, 177, 281 the Museum of the History of Science, Oxford. Figure 201 the Glasgow Museums and Art Galleries. Figure 214 Crown copyright, reproduced by gracious permission of H.M. the Queen. Figures 225, 227, 228, collection of Pierre Marly by courtesy *Plaisirs de France*. Figure 238 by permission of the Royal Society. Figure 239 Crown copyright, reproduced by permission the Royal Scottish Museum. Figure 262 Messrs Phillips Son & Neale. Figures 10, 11, 12, 22, 23, 59, 68, 70, 78, 99, 121, 127, 135, 146, 156, 185, 189, 256 have been photographed from private collections. All other photographs are by courtesy of Harriet Wynter Ltd.

INDEX